The Fleischmann Baths

DESIGN OF THE TURKISH BATH

By

J. J. COSGROVE

Author of

"PRINCIPLES AND PRACTICE OF PLUMBING"
"SEWAGE PURIFICATION AND DISPOSAL"
"HISTORY OF SANITATION"
"WROUGHT PIPE DRAINAGE SYSTEMS"
"PLUMBING ESTIMATES AND CONTRACTS"
"PLUMBING PLANS AND SPECIFICATIONS"

Books for Business
New York-Hong Kong

Design of the Turkish Bath

by
J. J. Cosgrove

ISBN: 0-89499-078-0

Copyright © 2001 by Books for Business

Reprinted from the 1913 edition

Books for Business
New York - Hong Kong
http://www.BusinessBooksInternational.com

PREFACE

♨ ♨

I N "Design of the Turkish Bath" it was the author's aim to present in as simple and comprehensive a manner as possible the method of constructing various types of Turkish Baths, their details and principles, together with descriptions of the materials and appliances best adapted to the purpose

This volume contains rules, tables and data for designing and proportioning Turkish Baths. The principles and data worked out by experiment and experience are so widely scattered through separate reports, public and private documents that they are not in available form. For this reason and owing to the present active interest in the subject it is hoped that the "Design of the Turkish Bath" will fill the want in the field of Sanitary Engineering literature.

These books, the author believes, cover very fully the various subjects, and contain such technical data as has long been needed by those interested in Plumbing and Sanitation.

J. J. COSGROVE

Philadelphia, Pa.
February 15th, 1913.

TABLE OF CONTENTS

LIST OF TABLES

LIST OF ILLUSTRATIONS

DESIGN OF THE TURKISH BATH

CHAPTER I

UNIVERSALITY OF THE SWEAT BATH

ROMAN BATHS

WHAT is known to people of the present day as the "Turkish Bath" is merely the highly evolved form of "sweat bath" practised for ages by both civilized and savage races. It would seem that human experience the world over has been the same, and widely scattered races of people, deducing certain truths from their personal experiences, early learned that if the pores of the skin were kept open, so that a copious flow of perspiration could be easily induced, beneficial results and improved health were sure to reward the effort. This led to communities establishing bath houses, crude or elaborate according to the wealth of the citizens or the advancement they had made in the building arts.

In Rome, no doubt, sweat baths reached the highest form of development. There the benefits of the sweat bath were known to the public at large, who availed themselves of the advantages

1

of the public baths, which were held in such high and well deserved esteem. Pliny, an early writer, credits the general good health of the Romans of that period to the frequent, almost daily, use of the hot room, and is authority for the statement that during the six hundred years of the public baths the Romans needed no medicine but the Thermæ.

Fig. 1
Ruins of Baths of Caracalla

Ruins of the Roman baths, see Figs. 1, 2 and 3, still exist, which show the former luxury and splendor of these establishments; and volumes have been written describing not only the general plans and details of the structures, but recounting, likewise, the habits and routine of the bathers, the use of the various compartments and the practices followed by different physicians.

From all that can be learned of the Roman practice of bathing it would seem that the bather first exposed himself to the temperature of the hot rooms, and, after going through the regular course of perspiration there, washed himself thoroughly with warm water, then closed the pores of the skin by the application of cold water, after which he was thoroughly massaged and the skin was anointed with oils, emollients, or skin foods.

A striking coincidence in this regard is that wherever the sweat house or hot bath has been established practically the same course of treat-

Fig. 2
Ruins, Thermæ of Caracalla

ment was followed, showing that human experience in all quarters of the globe taught the necessity of closing the pores after sweating and replacing the natural oils of the body removed by bathing.

SWEAT BATHS OF THE INDIANS

From the early Romans to the American Indians is a far cry, yet no more striking example of the widespread use of sweat baths could be cited than to tell the story of the American Indians' bathing establishments.

Every Indian village had its sweat house when the white man first came among them, so that the idea must have originated in America at the same time that it was in vogue in Rome, and could not have been borrowed from the white people. The sweat house of the Indians, as a rule, was built

Fig. 3

Mosaic from the Floor of the Baths of Caracalla

after the style of an ordinary Indian lodge, as may be seen in Fig. 4, with the exception that no opening was left near the top to serve as a smoke flue and ventilator. On the contrary, the framework of wood was covered with animal skins, sewed

together with tight seams to prevent the passage
of air from within or without. Inside of the sweat
house was built a rectangular heater made of stone
walls about six feet long, three feet wide, and
three feet deep. Every lodge in the village had a
woven willow basket or crib, which fitted the
heater on which it rested. The basket was made
large enough for a grown person to recline on

Fig 4
Frame for Toholua, or Sweat House of Indians

almost at full length and was supported in place
by cross poles which prevented the crib from drop-
ping down into the heater.

When a brave wanted to take a bath a squaw
built a fire outside of the sweat house and in this
fire heated a number of stones to a red heat.
These stones she then piled inside of the heater in
the sweat house, arranged the crib in place, put a

mat woven of wild sage and other aromatic herbs into the crib and the bath was ready for the brave. On the mat of wild herbs in the basket the Indian placed himself in a sitting posture and the squaw then poured cold water over the stones. The vapor, charged with the scent of the herbs, arising from the heater soon enveloped the bather, causing him to break out into a profuse perspiration, which continued as long as hot stones were placed in the

Fig. 5
Sweat House Used by Navajo Indians

heater. When the Indian had sweated to his heart's content he arose from the crib, ran to the river and plunged into the water. Emerging from the river he wrapped himself in a buffalo robe, ran to his lodge, where he was rubbed down and oiled with bear grease, then laid down, with his

6

feet to the fire, to sleep, wrapped in the buffalo robe with the fur turned inside.

The sweat house of the Indians was not always made of poles covered with animal skins. The style of lodge used for this purpose differed with the tribes and with their habits of living. Among the nomadic tribes, which were constantly moving from place to place, the tepee was made of animal skins as shown in the preceeding illustration, but among the more settled tribes sweat houses built of sods or sun-dried bricks, or dug out of the earth, were used instead. An illustration of a sweat house of this description belonging to the Navajo Indians may be seen in Fig. 5. This sweat house differs but little from caves used for a similar purpose by the natives on the far-away British Isles.

When the sweat bath of the Indians is compared with that of the Romans it is found to differ chiefly in its crudeness. So far as the routine observed is concerned there is practically no difference between the two. First, the brave subjected himself to the heat of the sweat room, which took the place of the tepid and thermæ chambers in the Roman baths. The next process, that of washing in hot water, was omitted, probably for want of suitable facilities. The third operation, consisting of closing the pores, was accomplished by plunging in the river, which the Indian used as the Roman did his plunge bath. The rubbing down or massage, of the Indian and subsequent anointing with oil in no wise differed

7

from that of the Roman, outside of the oils or unguents used.

There is one process, however, in which the bath of the Indian differs materially from that of the Roman, and it will be well here to note the difference, for upon this one process rests the chief distinction between what are now known respectfully as Turkish baths and Russian baths. In the Roman or Turkish baths, both as practised now and by the early Romans, the bather was subjected to a dry heat in the sweat rooms. Fire and hot gases circulated through flues in the walls of the old Roman baths and played beneath the floor of the hot room, so that the interiors of these rooms were like the inside of a baker's oven—dry, but extremely hot. In the Indian sweat house, on the other hand, the room was filled with vapor, so that instead of being a dry heat to which the bather was exposed it was distinctly a wet one. That is the chief difference between the Turkish bath and the Russian bath, to be described later on. The Turkish bath induces perspiration by means of dry heat, the Russian bath by means of vapor.

IRISH SWEAT BATHS

In Scotland and Ireland a sweat bath, very similar to that practised by the Indians, was likewise in vogue, and the fame of these baths became so great that it spread to Europe, where even at the present time in some localities sweat baths are known as Roman-Irish baths. The Irish bath houses, however, were very crude affairs, not at all to be compared with the baths of Rome, as may

8

be seen by a reference to Fig. 6, which shows a typical Irish dug-out sweat house. The baths were built close to the ocean or lake or running water, so that, like the Indians, the bather would have plenty of water for the operation and a pool in which to plunge when through with the heating

Fig. 6.
Old Sweat House in Ireland.

process. In the Irish sweat house a fire was built of peat on the floor and the fire was maintained until the walls were heated to a high temperature. The fire was then withdrawn, the bather entered and the hole by which he entered was closed to

9

keep the interior from cooling. Water was passed in to him as often as he wished it, to supply the moisture wrung from his body by the high temperature of the sweat house, or to wash himself, as the case may have required.

NATIVE RUSSIAN BATHS

The Russian bath, as practised in that country, differed but little from the bath of the Indians, as may be seen from the following description of the native Russian baths, taken from Kohl's "Russia":

"The passage from the door is divided into two, behind the check takers' port, one for the male and one for the female guests. We first enter an open space in which a set of men are sitting in a state of nudity on benches, those who have already bathed are dressing, while those who are going to undergo the process are taking off their clothes.

"Round this space or apartment are the doors leading to vapor rooms. The bather is ushered into them and finds himself in a room full of vapor, which is surrounded by a wooden platform rising in steps to near the roof of the room. The bather is made to lie down on one of the lower benches, and gradually to ascend to the higher and hotter ones. The first sensation on entering the room amounts almost to a feeling of suffocation. After you have been subjected for some time to a temperature which may rise to 145° the transpiration reaches its full activity and the sensation is very pleasant. The bath attendants come and flog you

10

with birch twigs, cover you with the lather of soap, afterward rub it off, and then hold you over a jet of ice-cold water. The shock is great but is followed by a pleasant feeling of great comfort and of alleviation of any rheumatic pains you may have had. In regular establishments you go after this and lie down on a bed for a time before issuing forth. But the Russians often dress in the open air, and instead of using the jet of cold water go and roll themselves in the snow."

In the numerous villages throughout the rural districts the same means employed by the early Indians are still used by the Russians for heating their bath houses or rooms. The bath house consists usually of two rooms, one for dressing and undressing, and the other the hot or vapor room. In the vapor room a heating stove is located, and on this stove or in its oven ordinary paving stones are heated very hot. Water is then poured over these stones to generate the vapor which fills the room.

In Japan the hot-water bath and sometimes a vapor bath are used, the object of either being to open the pores and cleanse the blood through the sweat glands in the skin.

It will thus be seen that people widely remote from one another—so remote, in fact, that they could not have adopted the practice from one certain place—each as the result of their own experience, proved to their own satisfaction the value of sweat baths. If these people were right in their deductions, as they certainly were, there must be good reasons why heat baths are beneficial to the health. Such, indeed, is the case, but in order to understand fully the benefits derived from the Turkish or Russian baths it will be necessary to point out first some of the chief characteristics of the skin and describe the functions of the numerous glands embedded therein.

CHAPTER II

HYGIENE OF THE BATH

FUNCTION OF THE BATH

THE SKIN

STRUCTURE of the Skin.—If a section of a piece of healthy skin be viewed through a microscope it will be found, as shown in Fig. 7, to be a many-functioned organ instead of a mere covering to protect the external surface of the body and support the internal organs. The outer, or scarf skin, will be seen to be made up of numerous scalelike cells of a horny consistency, which are being shed constantly and replaced by younger cells from below. Immediately beneath the outer, or scarf skin, is a layer of pigment cells. It is the difference in quantity and color of pigment deposited in these cells which gives to different races their characteristic colors.

Indenting this pigment layer at frequent intervals, so frequent, in fact, that fifty-seven thousand may be grouped together in one square inch of space, are little elevations, or centers, where blood vessels and nerve fibers terminate. Underneath the terminal of nerves and blood vessels is the true skin, which is sensitive. Beneath the true skin are fat cells embedded in fleshy tissue. Bedded deep in the true skin are a myriad of sweat glands with their little ducts leading through the various layers to the surface of the body. The roots of hairs are likewise embedded in

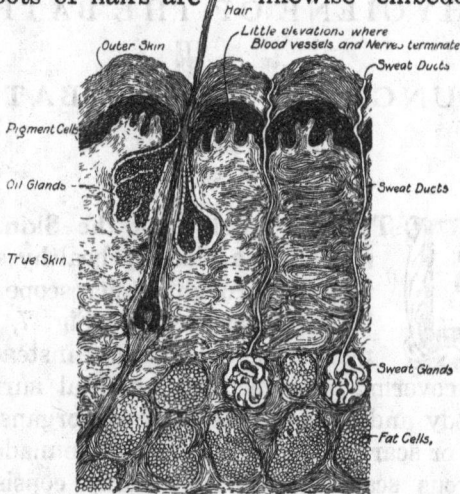

Fig. 7
Microscopic View of Healthy Skin

the true skin, and the stems extend out through oil ducts, which serve to conduct forth the oil secreted by the glands. A sectional view through an oil gland, oil duct and hair is shown, greatly enlarged, in Fig. 8. The root of the hair has its origin below the oil gland, where it communicates with a little nerve cell to give it the sensation of

14

touch, and with a blood vessel to supply it with nourishment.

Up through the center of the oil gland grows the hair and passes out to the surface of the skin, through the little oil duct, through which the gland pours forth its lubricating and softening fluid to the surface of the body. Oil glands, which are quite numerous, are scattered over the entire surface of the body, a gland and duct being found wherever there is a hair, as well as in many places where there are no hairs. The total number of hairs on the human body may be judged from the fact that there are estimated to be one hundred and twenty thousand hairs on a normal scalp.

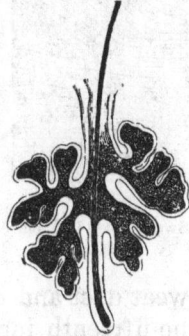

Fig. 8
Oil Gland, Oil Duct and Hair

More important, perhaps, than the oil glands and ducts are the numerous sweat glands embedded in the skin and the minute ducts which conduct their fluids out of the body. On the cheeks there are only about five hundred and fifty sweat glands per square inch; on the forehead there are twelve hundred glands per inch, and on the soles of the feet and palms of the hands, where they are the most numerous of any part of the body, they number as high as twenty-seven hundred per square inch. The little sweat tubes where they pierce the outer skin of the palms of the hands can be seen by a microscope dotting the curves, circles and whorls which make up the intricate patterns of the skin. These tubes can be

15

seen, greatly enlarged, in Fig. 9. According to careful computation there are about one million five hundred thousand sweat glands in the entire body, the total length of which has been variously estimated at from $2\frac{1}{2}$ to 28 lineal miles. When the

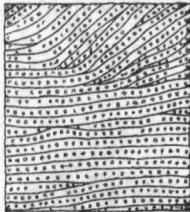

Fig. 9.
Sweat Pores in Palm of Hand.

number is conservatively estimated, however, the total length of perspiratory tubes will be found in the individual of average size to be less than 3 lineal miles. If the coils of the little sweat glands shown in Fig. 10 were unraveled and stretched out in a straight line it would be found that the combined length of sweat duct and coils of the gland measured about one-fifteenth inch. If the total number, one million five hundred thousand, were then placed end to end, they would be found to measure about 166,666 lineal inches, or 2.63 miles, in length.

It will be noticed that near the mouth of the tube the little sweat ducts are not straight, but take a tortuous or spiral course, like the coils of a spring or the spirals of a corkscrew.

Functions of the Skin.— When speaking of the skin, not only the scarf-skin and cutus, or true skin are meant, but likewise are included the various organs, glands, ducts, nerves and blood

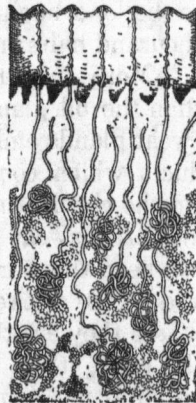

Fig. 10.
Sweat Glands and Ducts.

vessels embedded in the skin. The outer, or scarf-skin, being intended principally as a protective covering for the body, is admirably suited to that purpose. Made up as it is of numerous layers' of scalelike cells, composed of a horny substance, like the hoofs and horns of cattle, it is tough and will withstand an incredible amount of wear, while at the same time it is soft and pliable. This layer of the skin contains no blood vessels, so that it will not bleed when cut or scratched, and as it possesses no nerve centers is utterly devoid of feeling. The outer scales of this layer are constantly peeling off, exposing younger and more tender ones below, which are better suited to normal conditions. When, however, certain parts of the body are exposed to great wear—as, for example, the palms of the hands—a thick layer of horny scales or callus forms to protect those parts. The dead scales, which have become detached from the skin, or which hang like ragged particles to the surface, require periodical removing or they will putrefy on the body and perhaps clog some of the oil ducts or sweat pores.

To keep the outer skin, or scarf-skin, in a soft, smooth and pliable condition the various oil glands pour out their secretions of fatty matter, or oil, to lubricate the surface of the body.

When for any reason the action of the oil glands is interfered with or the ducts become obstructed no oil can be poured out onto the skin, which then becomes harsh and rough to the touch and, probably, chapped. The chapped hands most boys have in winter weather will readily present

17

themselves to most people as an example of such lack of activity of the oil glands, while the remedy, application of oil or grease to the rough parts, is simply doing artificially what nature does automatically, and without our being conscious of the fact. A turkish bath will likewise remedy the chapped condition by cleansing the surface of the skin and allowing the oil and sweat glands to perform their normal functions.

The pigment cells under the scarf-skin play a more important part in maintaining perfect health in the individual than would seem possible to the casual observer. The function of the pigment deposited in these cells is to prevent the penetration of harmful rays of light into the deeper tissues of the body. The action of the nervous organism in filling these pigment cells when necessary for the protection of an individual will be recalled in the "tanning" of dark-complexioned people when exposed to strong sunlight, and the sunburn, with subsequent tanning, of light-colored people when exposed to the direct rays of the sun.

So far as the bath is concerned the pigment cells are of no further interest, for the bath has no effect on the cells and the cells in no way interfere with the bath. In institutions where light baths are given, on the contrary, pigment, or the lack of pigment, in the cells would have to be considered, for not only has the pigment, or lack of pigment, a direct effect on the person subjected to the light, but, conversely, the intensity and color of light ought to be tempered to the color and amount of pigment in the cells.

The nerve centers which terminate in little elevations just beneath the layer of pigment cells impart to the skin the sense of touch, so that it will be sensitive and respond to influences of heat, cold, electricity or friction. The blood vessels, working in harmony with the nerve centers, drive the blood from the surface of the body at times, then, responding to some reaction, like heat or massage, draw it again to the surface and the skin takes on a healthy glow.

Of the numerous functions performed by a healthy skin there is, perhaps, no one gland, duct or vessel which plays a more important part than the little sweat glands. Through their pores these little glands, when in a healthy condition, throw off from the blood about two pints, or two pounds, of waste matter daily. In doing so the action of the glands is continuous, not intermittent, and the waste matter thrown off is generally in the form of invisable vapor, though when the body becomes warm from exercise or through excessive temperature of the atmosphere the vapor becomes visible in the form of perspiration. In this respect the sweat glands perform the glandular functions of the kidneys, and when the sweat glands refuse to perform their functions an extra burden is imposed upon the kidneys. This in itself would not be so serious a matter if the sweat glands and kidneys performed the same functions and were interchangeable in that respect, as they seem to be. As a matter of fact, however, they are not, for the function of the kidneys is to void fluids of an acid reaction, that being the normal reaction of urine, while the sweat glands excrete an alkaline solution, that being the normal reaction of perspiration. By exciting either of the organs they can be made apparently to perform the function of the other, but it is extremely doubtful if they actually do so. By exciting the sweat glands the fluid which would normally be voided by the kidneys can be carried off from the systems in the form of perspiration; but what becomes of the poisonous or deleterious acid constituents of that urine—are they carried off likewise by the sweat glands, which were not designed for the purpose? It would seem not; and, if not, such impurities will remain in the blood, accumulating with time, until the health of the individual becomes undermined. On the other hand, by exciting the kidneys much of the fluid required for perspiration may be carried off as urine. But, as the kidneys are intended only to filter out the acid wastes, what becomes of the alkaline matter which should be carried off through the sweat ducts? It would seem that each of the organs—the kidneys and the sweat glands—have separate and distinct functions to perform, and that one cannot successfully perform the functions of the other; even if it could it would seem an unwise policy by neglect of the person to require the kidneys to carry off daily the two pounds of matter ordinarily excreted by the sweat glands, or the sweat glands to carry off the waste ordinarily voided by the kidneys. That the kidneys or other organs cannot carry off the poisonous matters excreted by the sweat glands is proved by the fact that if the skin be varnished death will quickly result, due, no doubt, to the retention of some poisonous substance the nature and production of which are not understood; while the kidneys can be put out of service for a much longer time without fatal results.

To say that the sweat glands throw off two pounds of matter daily in the form of invisible perspiration is to state a fact without interpreting it. The bald statement lacks perspective to show its true size and proportion. A better understanding of this important function of the sweat glands can be had when it is known that the average amount of matter voided by the kidneys in twenty-four hours is approximately two and one-quarter pounds, or but slightly more than that thrown off by the sweat glands; while the solid fecal matter discharged by the average person in twenty-four hours is less than one-quarter pound and the vapor carried off by the lungs during the same period of time is approximately one pound.

There is another function of the skin in which the sweat glands play the most important part, and that is in maintaining a uniform and normal temperature of the body under all the variations of temperature to which individuals are exposed, so that, whether in the tropics, near the equator, or in the frozen north searching for the pole, the temperature of any individual would remain constant. The reason why temperature remains constant is because when exposed to a temperature greater than 98° Fahrenheit, or when the body temperature is raised by exercise, the little sweat ducts open wide their mouths and pour forth a

stream of moisture proportioned to the temperature or the exertion. The cooling effect of this
moisture evaporating from the surface of the body
keeps down the temperature, which otherwise
would rise to a dangerous or fatal degree. Owing
to this wise provision of nature man can live without the least discomfort when exposed to incredible
extremes of dry heat. For instance, a man can sit
with complete immunity in a hot-air bath raised to
a temperature sufficiently high to bake bread or
cook meat, and men have habitually sat without
the slightest inconvenience in the hot room of a
Turkish bath where the temperature was 240° Fahrenheit, or 28° above the boiling point of water.
Many instances could be cited of persons whose
occupations exposed them to temperatures of from
250° to 280° Fahrenheit for long periods of time,
and others who have endured temperatures of
360° Fahrenheit for shorter periods of time, but
these may be dispensed with and the case of Chabert alone mentioned. Chabert, who was known
to the public as the "Fire King," was in the habit
of entering an oven, the temperature of which was
from 400° to 600° Fahrenheit, and remaining there
a considerable length of time.

When exposed to a temperature of less than
normal body heat, on the other hand, the sweat
glands cease pouring forth their fluid, there is consequently no evaporation from the skin to reduce
the temperature of the body, and the internal combustion is required to furnish only enough heat to
replace that lost by radiation and convection.

In addition to the functions of the skin already enumerated the skin possesses also a respiratory function, giving off a small amount of carbon dioxide and taking in a small quantity of oxygen. In this respect the skin performs to a slight extent the function of the lungs. The skin can also absorb slight amounts of water and other fluids, but these last-named functions are of no great importance from the bathing standpoint.

Benefits of Bathing.—Bathing is beneficial to a person in a number of ways and affects, more or less, all of the glands, vessels, fibers and cells of which the skin is composed. These, in turn, react upon the larger and more important organs of the body, thereby toning up the entire system. As has already been mentioned, dead scales, or cells, are constantly being shed from the outer surface of the scarf-skin. Some of these scales, however, are not cast off at once, but cling to the younger cells beneath or hang in ragged fragments to the body until removed by friction or other mechanical means. These scales are dead organic matter and unless removed will putrefy on the person, besides, in many cases, interfering with or obstructing the oil ducts and sweat pores. The putrefactive bacteria engaged in breaking down the skin scales might contain, among their number, some harmful kinds which would cause illness or death if introduced to the body through a puncture of the skin. It will be readily seen, therefore, that a bath of any kind, taken regularly, will remove this skin as fast as it is shed, thereby preventing the mouths of oil ducts and sweat glands becoming clogged by

23

cast-off materials, will soften and remove the dead or dying skin which clings to the younger cells beneath, and will keep the skin physically, if not surgically, clean, thereby minimizing the danger from infection through a cut or bruise.

The beneficial influence of water not only upon the skin but likewise upon the nervous and circulating systems, as well as the internal organs of the bather, is due principally, to the temperatures of the water in which the bath is taken. Baths must, therefore, be considered according to their temperatures and modes of application, and the effect of hot and cold baths as well as vapor and hot-air baths will next be explained.

Cold Water Baths.—Generally speaking, it may be said that the effect of a cold bath is to close the pores of the skin, contract the capillaries, driving the blood from the surface of the body to the interior, leaving the skin white and bloodless. The functions of the skin being thereby temporarily arrested, the temperature of the blood rises from 2° to 4° Fahrenheit, which is equivalent to fever heat. Immediately after the bath, provided it has not been prolonged to an unreasonable extent, a reaction sets in, the sweat pores of the skin open, the capillaries expand and the overheated blood rushes back to the surface, bringing with it a healthy glow and a grateful feeling of warmth. So long as the healthy reaction can be induced, cold baths are beneficial to an individual, but if the skin remains white or turns blue the shock is too great and a warmer bath should be resorted to.

Baths in water from 65° to 55° Fahrenheit are considered cold, and anything below 55° Fahrenheit is considered very cold. Very cold baths cannot be borne long without ill effect, and baths which lower the temperature of the skin to 9° Fahrenheit may be endured for a very short time, but any further reduction of the temperature is liable to prove fatal. Baths in water from 80° to 65° Fahrenheit are considered cool baths.

When immersed in a bath of cold water the temperature of which is above 50° Fahrenheit there is a diminution of the temperature of the skin and tissues near the surface of the body and the temperature of the blood rises about 4° Fahrenheit. At the same time there is a slight shock experienced from the water, and if the cold is intense it induces a feeling of numbness in the skin, which becomes pale, due to contraction of the capillaries, which sends the blood to the internal organs.

As would be expected the cold bath likewise affects the central nervous system and the heart and lungs, as may be seen by the tremor of the limbs, the gasping for air, and the general depression which follows, due to the pulse beat becoming weaker. After the bath reaction sets in, bringing blood and warmth to the surface of the body. The colder the water and more powerful the depressing effect the quicker and more active will be the reaction, provided the individual is strong enough to withstand the shock.

Tepid Baths.—Any bath taken at a temperature between 80° and 92° Fahrenheit is a tepid bath. The effects of the tepid bath are not so numerous nor so far reaching as those of other temperatures, so that tepid baths can be borne for hours without ill results. The effects of the tepid bath are confined to the surface and do not reach the internal organs or nervous and circulating systems. There is no reaction whatever following the tepid bath, and the body and blood temperatures remain unchanged. On account of the absence of shock or stimulus of the internal organs tepid baths are best for people of weak constitutions or weak hearts. They are the least beneficial of all the baths, however, for those of strong, robust constitutions, as their beneficial influence is confined simply to cleansing the skin.

Warm Baths.—Temperatures from 92° to 98° Fahrenheit include those of the warm bath. There is no difference in effect between a warm bath and a hot bath; the only difference is in the degree of action and reaction excited. In the warm bath the effect is not confined to the surface, but is propagated to the internal organs, which causes an increased flow of blood to the surface and an increased frequency of the pulse beat. It seems likewise to stimulate slightly the building up or renewal of new tissue.

Hot Baths.—A bath at any temperature between 98° and 104° Fahrenheit is a hot bath. When a person is immersed in a hot bath a transfer of heat takes place from the warmer to the cooler medium—that is, from the hot water to the bather —while at the same time the evaporation becomes checked, and the combined effort of the transfer of heat and checking of evaporation increases the body temperature of the bather.

In the effort to keep down the body temperature the capillaries expand and blood rushes to the surface of the body, the skin becomes congested and the accumulated body heat finally bursts forth, causing a profuse perspiration, while at the same time the pulse beat increases and the respiration becomes quickened. It will thus be seen that in a hot bath the nervous and circulating systems become affected, which in turn react upon the internal organs. After the bath reaction sets in, the capillaries contract, all excess blood leaves the surface and the air at ordinary temperature feels cold to the skin.

Comparison of Hot and Cold Baths.—In comparing the effects of hot and cold baths upon normal individuals it might be stated as a general rule that the effects are opposite to each other; cold baths, on the one hand, tend to check perspiration, while hot baths favor it. It is believed, but not conclusively proven, that cold baths, by stimulating the internal organs, increase the reaction of the gastric and other fluids of the stomach and alimentary canal, while hot baths, on the other hand, tend rather to retard such activity. All baths, whether hot or cold, but particularly the latter, favor the secretion of urine.

Hot baths cause dilation of the capillaries and a rush of blood to the skin. When the stimulus of heat is withdrawn the capillaries contract and all excess blood flows away from the skin.

The cold bath, on the contrary, first contracts the capillaries and forces the blood to the interior, then when the reaction sets in the overheated blood flows back to the skin through the dilated capillaries.

A warm bath raises the temperature of the body by transferring heat to it and at the same time preventing evaporation and radiation of heat from the body. The cold bath reduces the temperature of the surface of the body by withdrawing heat from it but raises the temperature of the blood.

The hot bath draws the blood to the surface while the cold bath drives it to the interior. In either case there is increased oxidation, or waste of tissue, but with the warm bath there is less demand made upon the system, because oxidation depends chiefly upon increased heat, which in this case is supplied by the water. The reason that a hot bath seems refreshing to an exhausted person when he could not take a cold one may be that the heat supplied by the water helps the process of oxidation without any tax on the system.

The hot bath can be borne longer than a cold bath and a tepid bath can be borne longer than either.

Vapor Baths.—So far the effects of water baths only have been considered, no mention having been made of vapor or hot-air baths. It should be borne in mind, however, that the body bears changes of temperature of air, or even of vapor, much better than it does of water, because water being a better conductor of heat than either air or vapor brings more heat to the body or carries off a greater amount, as the case may be, than would air or vapor of equal temperature.

The vapor bath, on account of the less specific heat, does not act as quickly as water on the body, but once the action does begin it causes a profuse perspiration and acts powerfully in cleansing the skin. Vapor baths can be borne hotter than water baths, but not for so long, for the vapor being of higher temperature than the bather, and being charged with moisture, not only prevents evaporation from the skin but likewise radiation from the body. In consequence of this the temperature of the vapor bath, while far less than that of the hot-air bath, heats the blood considerably more, besides impeding the respiration by depositing moisture in the bronchial tubes. A vapor bath can be borne for a much longer time when the vapor is not inhaled. Ordinarily, however, when the vapor is inhaled a temperature of more than 125° Fahrenheit cannot be borne with comfort.

The effect of the vapor bath is much the same as that of the hot bath, with the one exception, perhaps, that it causes a more profuse perspiration, and in that one difference lies the greater value of the vapor bath. The vapor bath is the chief feature of the Russian bath, while the hot-air bath is the distinguishing feature of the Turkish bath.

Hot-air Baths.—Hot-air baths possess all the advantages of the hot-water bath, and some other advantages that hot-water baths, as well as other forms of bathing, do not possess. It is those latter qualities which give the hot air, or Turkish, bath its hygienic value.

One feature of the hot-air bath lies in the air which is inhaled. Unlike the air of a vapor bath it is not charged with vapor, so that moisture cannot be deposited in the bronchial tubes. Indeed, on account of the dryness of the air it increases instead of retards the evaporation from the lungs, while, being of a higher temperature than the body, there is no tax on the system heating the inspired air.

The greater benefit of the hot air bath, no doubt, is due to the profuse perspiration it induces without raising the temperature of the body over several degrees Fahrenheit. This profuse perspiration is beneficial in several ways. By exciting the glands it keeps them in a healthy and active condition; the fluids they pour forth flush the millions of little sweat ducts so that they will be perfectly clean, and last but not least, the fluid poured forth in the form of perspiration carries in solution waste products of the cell activity so that the blood is purified by the process.

It is in its function as a blood purifier, then, that the hot-air or Turkish bath differs most from ordinary baths, and it is in this function that the chief value lies, for any other form of bath will clean the person, stimulate the organs and excite the circulating systems.

Routine of the Turkish Bath.—The Turkish bath, as now practised, consists first of exposure of the naked body to dry, hot air until a profuse perspiration has been induced; next massage followed by a thorough scrubbing, with brush, soap and hot water; then a cooling shower to close the pores, which may or may not be followed by a cold plunge. The last stage consists of drying the body and resting.

Of the various operations undergone by a bather the exposure in the hot room is the most important and the massage is of the least benefit. When properly performed massage is of real benefit to a person, but slapping of the body with open palms to make a loud noise as commonly practised in Turkish bath houses, is of little or no value.

In its simplicity, then, the Turkish bath consists of exposure in the hot room, followed by kneading, rubbing, massage and a thorough washing, and this simple operation can be performed in the house as fully and with every bit as much benefit as in a public Turkish bath house, provided in addition to the ordinary bath room, there is a hot room or cabinet in which to induce perspiration. In the public bath room, however, numerous rooms and accessories are necessary in order to accommodate a number of patrons at one time and make the operation of bathing a luxury as well as a benefit.

Routine of the Russian Bath.—The Russian bath, as now practised, differs but slightly from that of the Turkish bath, the chief difference being in the use of a vapor room instead of the hot-air room used in the Turkish bath. After exposure in the vapor room until a perspiration has been induced the bather is put through the operations of kneading, rubbing, massage, scrubbing, shower, plunge, drying and rest as in the former case.

31

In most Turkish bath establishments a vapor room is provided in addition to the hot-air rooms, so that patrons may take either a Turkish or a Russian bath, or combine the two.

Vapor and hot-air baths are so beneficial to a person that it is to be regretted that provision is not more frequently made for them in the home. The matter would be simple to effect, for a small room or cabinet could easily be provided and fitted with means for heating and ventilating, so that the inmates of the house could have both the benefit and the luxury of a Turkish bath.

CHAPTER III

LAYOUT OF A TURKISH BATH

ARRANGEMENT AND RELATION OF THE VARIOUS PARTS

CONVENTIONAL Floor Plan.—
The conventional layout of a simple
little Turkish-bath establishment is
shown in Fig. 11. In a large bathing
establishment many features would
be included which are not shown here. For instance, a barber shop, lounging room, pedicure
booth, hydrotherapeutic department, vapor room,
sun parlor, light baths and massage room might be
included, or any one or more of them. Furthermore, instead of combining the shower, plunge and
shampoo, or laving rooms, all in one, as was done
in the present instance, separate rooms or compartments would or could be provided for each.

The example here illustrated is given because
of its simplicity and because it shows at the same
time all the necessary elements of the bath. Dressing rooms containing a cot, bed or couch are nec-

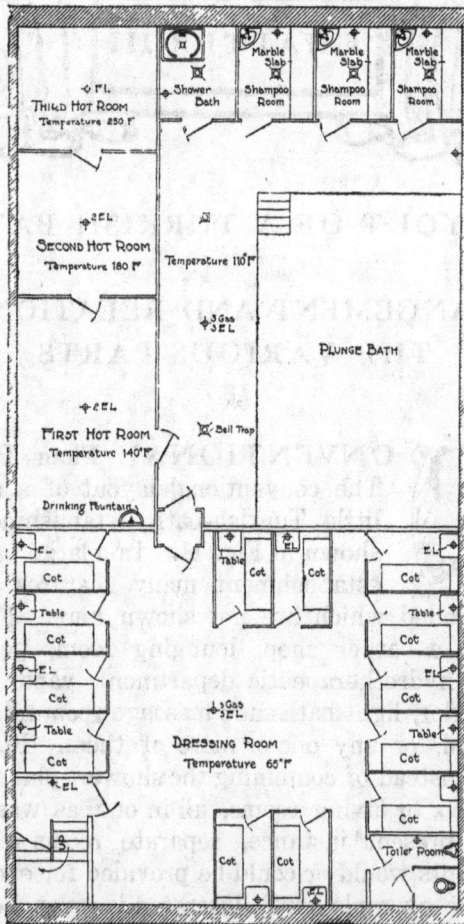

Fig. 11
Conventional Floor Plan of Bath

essary in which to dress and undress, as well as to rest in and possibly sleep after the bath. Besides the bed or couch each room should contain a chair and dressing table, with mirror and hooks on which to hang clothes. Each door should be provided with a Yale or other lock which cannot be easily picked, and the top of the rooms should be covered with a wire netting unless the walls extend to near the ceiling. At some convenient place in the establishment toilet facilities for the bathers should be provided. The number of fixtures need not be large, one water closet and one urinal being sufficient for each fifteen persons that can be accommodated at one time.

From the dressing room the bather passes into that part of the establishment where the actul process of bathing is undergone. There are first the the hot rooms, next the shampoo compartments, then the needle shower and spray bath and, finally, the plunge, each of which will be considered separately

THE HOT ROOMS

Number of Hot Rooms.—In the bathing establishment of the ancient Romans the bather passed in successive stages through three hot rooms, known respectively as the *tepidarium*, or warm room, *calidarium*, or hot room, and the *loconium*, or hottest room. By thus accustoming himself to the temperature of one room before passing to the next hotter one he was able to endure a much higher temperature and, consequently, derived much greater benefit from the bath than

35

if he were exposed only to the heat of one room and the temperature in that room was as hot as he could endure.

This ancient practice of the Romans has not been improved upon to this day, and the best results will be obtained in Turkish bath design by arranging three hot rooms *en suite*, as shown in the example illustrated.

The human system can accustom itself to extremes of temperature and pressure if the changes leading up to the extremes are easy and gradual, not sudden and severe. Everybody is familiar with the danger of passing suddenly from ordinary atmospheric pressure into compressed air of two or more atmospheres, or of emerging suddenly from the higher to the lower pressure, while at the same time the transition can be made without danger if the change is gradual, so that the body can become accustomed and adjust itself to one pressure before being exposed to another higher or lower one.

In the same way, no doubt, the body is put under a severe strain by exposing it suddenly in a room heated to an extremely high temperature, while, on the other hand, if the high temperature be reached gradually, by passing from one hot room to a hotter one, until the maximum temperature is finally reached, the effect will be beneficial. The Indians seem to have understood this fact as well as the Romans; they accomplished the same result by entering their sweat house when it was but slightly above the normal temperature and remaining until it had reached the maximum. In spite of the benefit derived from a gradual exposure to the highest temperature the tendency of the times in Turkish bath design is to provide but one, or sometimes two, hot rooms for the bathers. Two hot rooms are not so bad, but the cutting down of the number of hot rooms to one cannot be commended unless the gradual heating is achieved in some other way. For instance, if the room has a high ceiling the difference in temperature between the floor level and the ceiling would be great, and different temperatures could be attained by the bathers if balconies were placed at different elevations in the room.

But there is a different way of looking at the matter. People differ greatly in the amount and degree of heat they can stand. Some have reached their limit at 140° Fahrenheit while others delight in the higher temperature of 250° Fahrenheit, and if the bath is to be considered as a pleasure and luxury the different temperatures desired by the various patrons should be maintained, and this, of course, is impossible when there is only one hot room. Again, Turkish baths are often used for reducing flesh and as remedial measures in case of sickness, and for such cases the higher-temperature rooms will be required even though most patrons desire only the low-temperature room.

Temperature of Hot Rooms.—The temperatures to be maintained in the various hot rooms of a Turkish bath have been well established by usage. In the first hot room the temperature may vary from 130° to 150° Fahrenheit with an average of 140° Fahrenheit. In the second hot room the temperature may vary between 170° and 190° Fahrenheit with a fair average of 180° Fahrenheit; while in the hottest room the temperature will average 250° Fahrenheit and may vary from 240° to 260° Fahrenheit. In bath establishments where there is but one hot room it would seem that a temperature equal to the second hot room would be about right, while in establishments having two hot rooms temperatures similar to the first and second hot rooms would prove satisfactory. Of course, in fitting up such a room sufficient radiation would be supplied so that hotter temperatures could be maintained if necessary, and the radiation would be so controlled that the various units could be cut off or turned on at will to regulate the amount of heat by adding to or taking from the heating surface.

In speaking of the temperature of the hot rooms it is assumed that the thermometer readings are taken at an elevation of 6 feet 6 inches above the floor line.

Not only the temperature of the hot rooms, but, likewise, the temperature of the other parts of the bathing establishment must be considered when designing a Turkish bath. In the shampoo compartments, likewise the shower room and plunge bath, the temperature of the air should be kept at about 110° Fahrenheit. Owing to the lack of clothing and the rapid evaporation from the body there would be a feeling of chilliness in these rooms if they were maintained at lower temperatures. In the dressing rooms, on the other hand, the bathers are provided with sufficient clothing to keep them warm, and in order that they get the greatest amount of good from the bath, the temperature there should be kept at about 65° Fahrenheit and there should be a constant change of pure, fresh air.

Size and Proportion of Hot Rooms.—The size of the hot rooms for a Turkish bath will generally depend upon the size of the building and the magnitude of the establishment. Seldom will they be made so small that the amount of space to be allotted to each patron will have to be considered, for Turkish baths are built more as a luxury than as a convenience or necessity, consequently ample room will be provided for all compartments. Nevertheless, there will be cases where that problem arises, and it will be well to consider just what amount of space will be necessary for each person. In the first place, a certain amount of space will have to be reserved for passages, and, after allowing for such aisles, a fair allowance for each person would be a floor space occupying a rectangle 6 feet long by 4 feet wide. Such an allowance would give approximately 2 feet between chairs and would allow about 18 inches over the average length of reclining chair, such as are generally used in the hot rooms of bathing establishments. Next in importance to the amount of space to be allotted to each patron is the proportion of the various hot rooms to one another. Experience shows that of the numerous patrons of a bathing establishment only a certain percentage of them will pass into the hotter rooms.

All of the patrons, however, enter the first hot room, so that this compartment should occupy from one-half to three-fifths of the total space allotted to the hot rooms. Of the remaining space a little over one-half, or about three-fifths, should be allotted to the second hot room, and the balance of the space to the hottest compartment. The hottest compartment, it would seem unnecessary to point out, is made the smallest because fewer people will reach this room than will expose themselves to the heat of the other rooms.

Drainage for Hot Rooms.—As a rule it is better not to place a floor drain in any of the three hot rooms of a Turkish bath. Owing to the dry heat in these compartments and the absence of water, except when cleaning and scrubbing the rooms, the traps would be liable to become emptied by evaporation, thus allowing drain air to enter and vitiate the air within the rooms, which should be kept as pure and fresh as possible. Instead of placing a floor drain within the sweat rooms the better plan is to locate one just outside, as shown in the illustration, then give the entire floor of the three hot rooms a slight slope towards this drain.

Water Supply for Hot Rooms.—The only supply of water ever piped to the hot rooms of a Turkish bath is a line of filtered water to a drinking fountain, located in the first hot room. On account of the amount of moisture lost during the routine of the bath, particularly when in the hot rooms, it is advisable to drink as much water as possible, so that transpired perspiration will not take place at the cost of the living cells, which might be robbed of their much-needed fluids. On account of the faucet and some of the piping being located within the hot room the water from such a drinking fountain will be found slightly warm, particularly if not allowed to run for a short time before drinking. Warm water, however, promotes perspiration, while cold water tends to check it, so that those who like warm water or, at worst, can drink it, will be benefited by so doing.

Ventilation of the Hot Rooms.—The ventilation of the hot rooms is of the utmost importance to the bathers. The methods of ventilation are taken up in a subsequent chapter and all that is necessary here is to point out, generally, what is good and what is bad in this respect. It will be remembered that the object of the bath is to gradually raise the temperature of the body until all the sweat ducts open wide and pour forth the stored up impurities from the blood. The greater comfort with which this can be accomplished, the better it will be for the bather in every way. Lack of air is a discomfort which would defeat much of the object of the bath, and a supply of unsuitable air cannot be considered in the nature of ventilation. It is a well-known fact that heating air to a high temperature deprives it of some of its most refreshing qualities. Note, for instance, the invigorating cool air of early Spring, Autumn, or the sea shore, as compared with the enervating hot air of July and August. There is the same quantity of air available at all times, but the quality differs greatly under different conditions. The benefits of the hot bath would, perhaps, be greatly enhanced if the bather could draw in his lungs copious drafts of fresh cool air from the outside while sitting exposed

41

to the radiant heat of the hot room. Such a condition would simulate the condition of an early Spring morning when seated in a protected nook of a garden exposed to the sun enjoying the invigorating glow from a sun bath—radiant heat—while breathing in pure cool air of Spring.

Perhaps sometime we will be able to provide similar conditions in the hot rooms of a Turkish bath. A hood or vent duct in which the heads are immersed in fresh cool air, while the bodies are exposed to the radiant heat of the steam coils, or a supply of fresh cool air supplied in some other way to the nostrils, would seem to add greatly to the comfort of the bathers in the hot rooms, and if to the comfort, then to the benefit. Some experiments would have to be made before such a method would be practicable, but the benefits that would be derived warrant keeping the possibilities in mind.

CHAPTER IV

INSULATING TURKISH BATH ROOMS

INSULATION of Hot Rooms.—In order that the hot rooms of a Turkish bath may be economically maintained at their proper temperatures it is absolutely necessary to thoroughly insulate the walls and ceilings so that heat will not be constantly radiated to the surrounding atmosphere or penetrate to other parts of the building. Usually a Turkish bath establishment is but one of many industries grouped together under one roof, and if heat from the bath rooms were permitted to escape to other parts of the building it would not only entail an extra and unnecessary expense on the proprietor of the bathing establishment but would, furthermore, prove extremely annoying and, perhaps, expensive to the rest of the tenants.

When selecting insulating materials it should be borne in mind that dead-air space is among the very best of heat insulators or retardants. If the air has room for circulation, however, the insulating effect will be destroyed. That is why materials like mineral wool, which of themselves are

43

heat conductors, will, when packed in an air space, prove effective for insulating walls or ceilings. The mineral wool when packed into the space forms a multitude of cells which contain air, but, owing to the cellular formation of the mass, the air is dead and cannot circulate. Hollow tiles likewise prove good insulators. In the case of hollow tiles, unlike spaces filled with mineral wool, there is local circulation, but of such a sluggish and restricted nature that the air is practically dead.

Besides mineral wool any fibrous or granular material which will not pack too tight will prove satisfactory for filling purposes. Among the materials successfully used may be mentioned coal cinders, hair felt, granulated cork, charcoal and sawdust. Materials like granulated cork and charcoal rank highest for the purpose for the reason that in themselves they are not good conductors of heat, while materials like mineral wool and cinders are, to a limited extent. On the other hand, cinders and mineral wool are fireproof, while some of the better insulating materials are not, and in designing an establishment all these facts must be taken into consideration.

Concrete is a much better insulator of heat than is generally supposed. This fact was recently demonstrated very forcibly at Boston, Mass., where, for five hours, a floor 5 inches thick was subjected to an intensely hot fire, varying in temperature from 1,700° to 2,000° Fahrenheit. During the entire period of the experiment there was not a moment when the hand could not have been held on the top surface of the concrete without fear of being burned.

Insulating Fireproof Walls.—Certain methods for insulating walls, floors and ceilings against the loss of heat have been adopted in practice for the reasons that they are simple, effective and applicable to all conditions. The method commonly adopted for insulating walls of stone, brick or concrete is shown in Fig. 12. The inside of the wall to be insulated is first given two good coats of pitch, tar, asphaltum or some like elastic and moisture-proof material to prevent dampness striking into the wall and to keep down as low as possible the loss of heat due to winds. Inside of the coat of pitch is laid a wall of hollow tile to provide a partial dead-air space. Four inches inside of the first wall of hollow tile is laid another, 3 inches thick, and between the two courses of tile the space is filled with mineral wool. The inner surface of the 3-inch hollow tiles can then be plastered with cement mortar for a finish, or, if a better finish is desired, the surface can be covered with marble or tiles. Instead

Brick Wall
2 Coats of Pitch
2" Hollow tile
4" Mineral Wool
3" Hollow Tile.
Cemet Plaster or Tile.

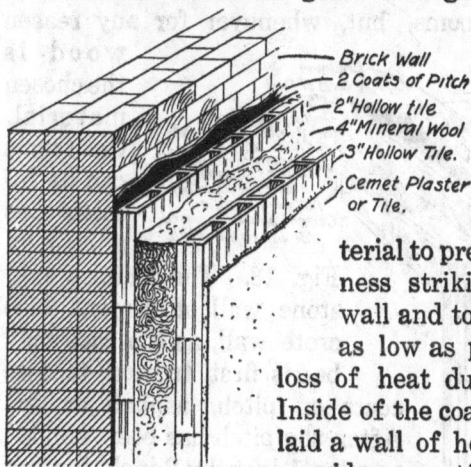

Fig. 12.
Fireproof Insulation for Masonry Walls.

45

of the inner course of hollow tiles, glazed bricks may be laid, and the space between the bricks and the first course of hollow tile filled in with mineral wool or some other filling material, as previously explained. Further, wherever necessary or desirable to do so, one course of hollow brick can be omitted and the wall otherwise insulated just as described.

Wooden Insulation for Brickwork.—Wood is not so desirable a material as hollow tile for insulating hot rooms, but, whenever for any reason wood is the chosen material, the rooms can be insulated as shown in Fig. 13. The brickwork, stone wall or cement concrete wall, as the case may be, is first treated to two coats of pitch, asphalt or tar. After the pitch has been applied regular 1 inch by 2 inch furring strips are nailed to the wall at distances of 16 inches from center to center. A horizontal course of ⅞-inch sheathing is nailed to the furring strips, a double layer of good building paper tacked to the sheathing, and another course of sheathing, laid diagonally, is placed on top of the building paper and nailed to the furring strips through the first course of sheath-

Brickwall
2 Coats of Pitch
1ʺx2ʺStrips 16ʺc-c
Double 1ʺSheathing with 2 layers of paper betw
2x4ʺStuds 16ʺc-c filled between with mineral wool
Double 1ʺSheathing with 2 layers of paper betw.
1ʺx2ʺStrips 16ʺc-c
Cement plaster or glazed tile on exp metal lath.

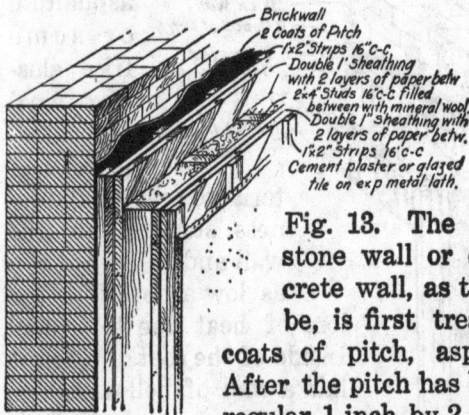

Fig. 13
Wooden Insulation for Masonry Walls

ing. Regular stock, 2x4, studding are then set up on 16-inch centers and nailed to the sheathing and the face of the studding covered with a double course of sheathing with building paper between, as in the first instance. The space between the studding is then filled with mineral wool, or some other heat insulator. Furring strips are next nailed to the surface of the last course of sheathing, and expanded metal lathing fastened to the strips, so that the surface is ready for a three-coat layer of cement plaster or for a finish of glazed tile. If the masonry walls are pretty thick, or are otherwise well insulated, the inner layer of wooden sheathing may be omitted, and the expanded metal lath nailed to the face of the 2 x 4 studding.

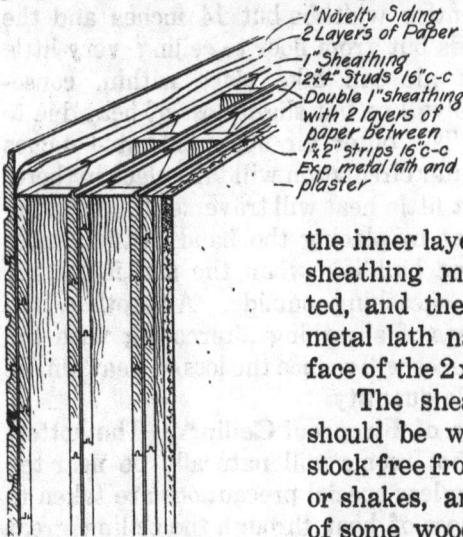

The sheathing used should be well matched stock free from knot holes or shakes, and should be of some wood which will not twist and warp under the action of heat.

Fig 14

Insulation for Frame Walls

Labels on figure:
- 1" Novelty Siding
- 2 Layers of Paper
- 1" Sheathing
- 2x4" Studs 16"c-c
- Double 1" sheathing with 2 layers of paper between
- 1x2" strips 16"c-c
- Exp metal lath and plaster

Insulation of Walls in Frame Buildings.—The method of insulating walls in frame buildings may be seen in Fig. 14.

47

This method is similar to the wooden insulation for brick walls just described, with the one exception that 2 x 4 studding take the place of the outer 1 x 2 furring strips, and a double course of sheathing with building paper between takes the place of the brick wall and forms the outer surface of the building or wall. This method of insulation gives three air spaces throughout the depth of the wall. In one of the spaces the cells between the studding are filled with mineral wool so that in these the air is dead. One of the air spaces is but 1 inch deep, and as the entire width is but 14 inches and the height extends but from floor to ceiling very little circulation of air can take place within, consequently there will be but slight loss of heat due to convection. The outer air space is but 4 inches deep and greater circulation will take place in there. However, but little heat will traverse these spaces, as may be seen by placing the hand on the plaster within a frame building when the sun is beating down on the sheathing outside. Altogether, the various courses of sheathing alternating with the several air spaces will reduce the loss of heat almost to a negligable quantity.

Insulation of Fireproof Ceilings.—The hottest part of the hot rooms will naturally be near the ceiling, and unless special precautions are taken to prevent the loss of heat through the ceiling great inconvenience will be caused the tenants above, and much valuable heat will be lost. A method of insulating a flat-arch ceiling is shown in Fig. 15. The floor tiles themselves, being double-space hollow patterns, will go a long way toward checking the radiation of heat at this point, particularly when covered on the underside with a coating of cement plaster or a surface of glazed tiles. On top of the floor tiles is a deep bed of cinders from 6 to 10 inches in depth, and embeded in these cinders are heavy wooden sleepers to support the floor above. The flooring consists of two heavy layers of tongue and grooved material, free from knot holes or shakes, laid up tight with a double layer of good building paper between. The under floor is laid diagonal and the upper floor at right angles to the sleepers, so long joints in the two floors cannot coincide, and, furthermore, so the upper floor can better prevent warping of the under flooring strips. If necessary or desirable any other kind of flooring may be laid instead of a

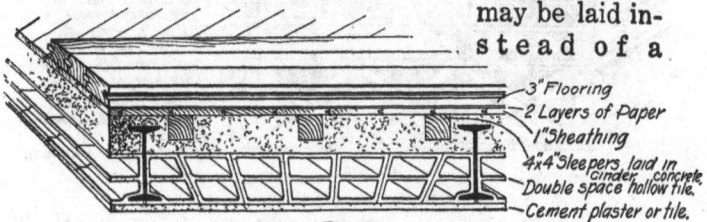

Fig. 15
Insulation for Fireproof Ceilings

wooden one. A cement floor, tile floor, asphalt or any other material will answer just as well and prove equally effective. There are other forms of arch besides the flat arch and other types of ceiling tiles besides those shown in the illustration. The method of insulation, however, would apply equally to all, with whatever modifications good judgment would dictate.

Insulation of Wooden Ceilings.—In frame buildings special precautions must be taken to prevent the radiation of heat which would take place through an ordinary plaster ceiling and single course of flooring above. Ceilings in frame buildings can be satisfactorily insulated, as shown in Fig. 16. The under edges of the floor joists may be covered with expanded metal and then plastered with cement mortar or covered with tiles, or, if wooden treatment is to be carried throughout the construction, the under surface of the joists can be sealed with a double course of matched ceiling boards, having a double layer of building paper between. On top of the floor or ceiling joists

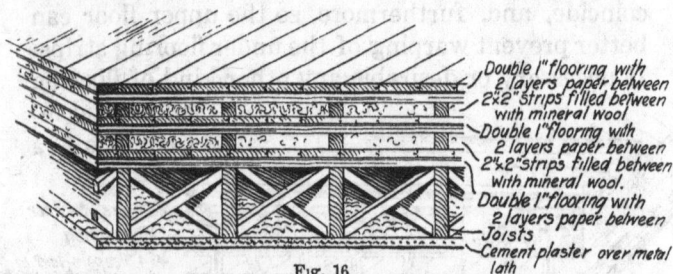

Double 1" flooring with 2 layers paper between
2"x2" strips filled between with mineral wool
Double 1" flooring with 2 layers paper between
2"x2" strips filled between with mineral wool.
Double 1" flooring with 2 layers paper between
Joists
Cement plaster over metal lath

Fig. 16

Insulation for Wooden Ceilings

another double course of matched flooring is laid with two layers of good building paper between. On top of this double flooring 2-inch by 2-inch strips are nailed at intervals of 16 inches between centers; the spaces between these strips are then filled flush to the top with mineral wool, and another double course of flooring is laid with building paper in between the courses. On top of this flooring 2-inch by 2-inch strips are likewise nailed, the spaces between filled with mineral wool, as in the first instance, and another double course of flooring is laid on top, with building paper placed between the courses. The top boards in this case would, of course, be the finished floor, or if cement or a tile floor is to be laid the top rows of 2 inch by 2-inch strips may be omitted and the spaces filled with cinders, on top of which a rough bed of concrete should be laid to provide foundation for the finish course.

When laying the double floors or double courses of ceiling boards, as the case may be, one course should be run diagonally and the other at right angles to the studding, joists or furring strips.

Insulation of Ground Floor.—By the ground floor is meant the floor of the cellar or basement of a building which is in direct contract with the earth. There is more or less ground water stored in the soil, and from this subterranean reservoir a greater or less amount of moisture is drawn by capillarity toward the surface of the ground. Water has a great capacity for heat, and if soil moisture were allowed to rise to the ground floor of a building in which a Turkish bath hot room was located much valuable heat would pass into the surrounding earth. To prevent such loss of heat, as well as to keep the ground floor dry, it may be insulated as shown in Fig. 17. The cellar bottom must first be dug out from 24 to 30 inches deeper than the line of the finish floor. The bottom of this pit should then be filled to a depth of from 6 to 10 inches with rocks or crushed stones. On top of the stones would next be placed a layer of good clean cinders, from 12 to 16 inches deep; embedded in these cinders, but well supported at suitable intervals, should be placed 4 x 4 inch sleepers, spaced 16 inches from center to center. On top of the sleepers there would be a regulation double floor with two layers of building paper between the courses. On top of the flooring there would be more 4 x 4 inch sleepers with the spaces between filled with mineral wool, and on top of the sleepers may be laid a regulation wooden floor consisting of two courses with building paper between, or a cement, asphalt or tile floor may be laid. Of course, inside of the hot rooms an asphalt floor

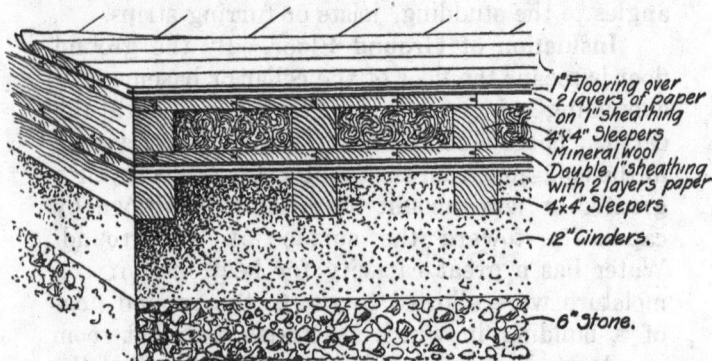

Fig. 17
Insulation for Ground Floor

would not be desirable on account of the odor from hot asphalt and the fact that the flooring would become soft from the heat. It might be well to add that in case the floor is to be of tile or cement that the woodwork would be left out entirely and the concrete and cinders depended upon for the necessary insulation.

Insulation of Windows.—When windows or skylights open into any of the rooms of a Turkish bath which are heated above ordinary room temperatures great care must be exercised to prevent the loss of heat through the glass. The ordinary window sash with single-thick or double-thick glass would never do for this purpose. Sash and frames of special construction must be provided. The method of insulating windows is shown in Fig. 18. Ventilation to the rooms is provided for, independent of the windows, so the sash need not be made to raise and lower like those of ordinary windows. That being true, the sash are made to fit the entire window opening and are set in place so that the

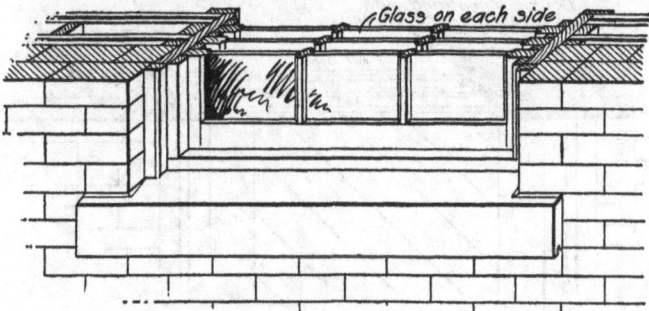

Fig. 18
Insulation for Windows

leakage of air around the stops is reduced to the minimum. It will be noticed that instead of one sash there are three sash in this frame, and each sash is glazed with two panes of glass, so that all told there are six thicknesses of glass and five air spaces for the heat to pass through. If the window exposure of a Turkish bath is at the south or in a sheltered part of the east, or if the country is a comparatively mild one, two sash, with four thicknesses of glass and three air spaces, would probably be sufficient. In glazing windows for Turkish bath rooms it is customary not to use putty and window tacks, as is commonly done in glazing, but to use adjustable stops with felt or chamois linings, so the glass will have room and freedom to expand and contract without opening crevices for the leakage of air. To avoid breakage the panes of glass should not be very large, and rolled glass should be used in preference to ordinary window glass, as rolled glass is tougher and less liable to break.

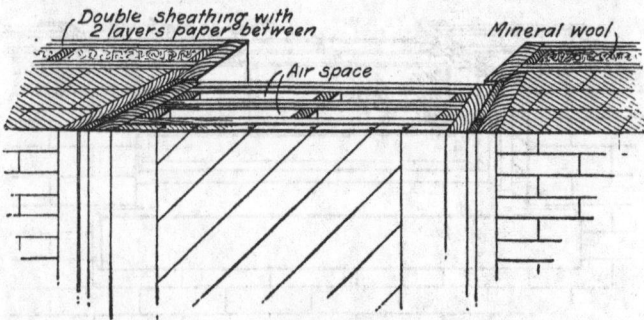

Fig. 19

Insulation for Doors

Insulation of Doors.—Wherever a passageway leads from one of the hot rooms of a Turkish bath to a cool room—that is, one maintained at the ordinary temperature of living rooms—double doors with a lobby between will be found advisable to prevent the transmission of heat from the hot to the cooler rooms. In other parts of the bathing establishment the opening from one room to another of higher temperature may be closed with

54

a single door which is well insulated against the passage of heat.

An insulated door is shown in Fig. 19. This door has three double layers of boards forming a double air space, and three layers of good building paper are placed between each double thickness of board.

In the illustration an ordinary door is shown without attempt at ornamentation. When, however, the bathing establishment is to be elaborately finished the doors may be surfaced with the wood used for the interior trim and may be paneled to correspond with any detail.

If for any reason it is found necessary or desirable to place glass in the doors it can be done by observing the precautions necessary for window insulation. It might be well to have the upper part of the door leading to the first hot room made of glass, so that bath attendants from the outside can see if everything is all right with the bathers without having to open the door, as would be necessary if the glass were omitted.

Patented Heat Insulating Materials.—In the foregoing pages the methods of insulating against the loss or leakage of heat, and the doors and windows used for this purpose, have of necessity been stock goods obtainable in any locality, and easily assembled from goods at hand. The principles outlined are the principles which must be observed in any kind of insulating methods or materials, whether ordinary methods or patented ones. However, there are certain patented heat insulating materials now in common use which have been made with the one object of preventing heat loss. Likewise there are patented doors and windows, and in many cases insulating materials, doors and windows which are manufactured for the purpose but not patented, and often those will prove not only cheaper but better than the ordinary methods outlined, the difference depending on the method of construction and the location of the building.

The practice of heat insulation has been brought to its highest state of perfection in the ice-making and refrigeration industry, where heat loss must be reduced to the minimum if an economical operating plant is expected. When taking up the design of the Turkish bath, therefor, the manufactured materials for insulating buildings, walls, floors, ceilings, windows and doors can be found advertised and described in publications devoted to ice-making and cold storage, to which it will be advisable to refer.

CHAPTER V

HEATING TURKISH BATH ROOMS

METHODS of Heating the Sweat Rooms.—Up to the present time the Turkish bath has been referred to in this work as a hot-air bath. As a matter of fact, however, the Turkish bath is a heat bath pure and simple, the temperature alone being depended upon to bring about the beneficial results sought in the sweat rooms, regardless of the manner in which the heat is applied. For convenience the term hot-air bath will continue to be used, although it must be understood that this term is meant simply to distinguish hot baths in air from hot baths in water.

So far as heating of the sweat rooms of a Turkish bath is concerned, in the best method used, the direct radiation method, the air is heated only incidentally, not with a view of transmitting the heat from the air to the bather. In the hot-air method, on the other hand, air is used as a vehicle of heat, and those baths are not only heat baths, but are rightly named hot-air baths. It should be born in mind, however, that the object of a Turkish bath is to apply *heat* to the body, not necessarily air or hot air.

There are two systems now in use for heating Turkish sweat rooms. The first, the hot-air method, is similar to heating an ordinary dwelling by means of a furnace, only the rooms must be heated to a much higher temperature. In the second method direct radiation is employed, which may be likened to the heat from a muffle kiln where earthenware is being fired. The second method, heating by direct radiation, was the means employed by the ancient Romans for heating their bathing establishments, and is the method, greatly modified but not improved, which is followed at the present time in the United States. In Great Britain, on the other hand, the hot-air method is the one commonly followed.

In Turkish bath rooms, heated by radiation, the heat is transmitted direct to the bather from the walls, floor and ceiling of the rooms which are heated to a high temperature from outside, or from a system of steam pipes or radiators placed within the rooms. In this respect the Turkish hot rooms are very similiar to the bake oven of an ordinary kitchen range, which is heated from outside and in which only an occasional change of air takes place. In the sweat rooms, however, provision is made for ventilation, which will be explained later. The old Roman baths were constructed with the furnace underneath the sweat rooms and so designed that the flames and hot gases, after playing upon the ceiling, which formed the floor of the hot rooms, passed up through hollow tiles or flues formed in the walls of the sweat rooms, so that radiant heat flowed to the bather from the walls, the floor and the ceiling of the rooms.

In the British hot-air method of heating, a steady stream of air of the temperature desired is passed through the hot rooms, the volume and velocity being sufficient to maintain the temperatures of the several rooms without causing draught.

Each method of heating has its advantages and disadvantages, but it seems to be pretty generally conceded that radiant heat is the most beneficial form of heat and for bathing purposes better than hot air. It would seem at first thought that heat is a condition or vibration which is the same no matter from what source it is derived or in what form the energy is available. Such is not the case, however, as may be seen by a comparison of the two methods of heating under consideration.

Take, for example, first, the interior of a hot room heated by radiation from the walls, floor and ceiling, and bear in mind that heat always flows from the hotter to the cooler substance and that radiant heat passes through air without heating it. In that case, the walls, floor and ceiling being approximately of the same temperature, there would be but little transfer of heat from one point to another within the room; on the contrary, the heat would be, in a nature, "bound," or at rest. Let a bather then step into the room which, assume, is heated to a temperature of 200° Fahrenheit. The moment he does so the radiant heat from all parts of the enclosure bombards him with a multitude of heat waves which strike him from every quarter and play upon him with an intensity equal to the difference between his own body temperature and the temperature of the walls. In addition to the heat waves, it is believed, but not fully demonstrated, that electro-thermal currents also pass from the waves to the bather through the dry air of the room. In short, all of the excess heat above the body temperature contained in the walls, floor and ceiling of the rooms focus their rays upon the bather and concentrate their combined force upon his person, giving him its undivided attention.

In the second case, when the rooms are heated with hot air, all heat contained in the walls, floor, ceiling and air of the rooms must be supplied from the current of air pouring in at the register. That being the case, as soon as air enters the room the heat imparted to it by the furnace is divided, part flowing to the bather, part supplying the heat lost from walls, floor and ceiling, and part passing out through the ventilator register. Instead of the heat being concentrated on the bather it is scattered and much of its beneficial effect is lost.

Besides the concentration of heat waves in the one case and dissipation of them in the other, there is a real benefit derived from radiant heat which cannot be secured from heat by convection. What the quality of radiant heat is, which makes it so superior to other forms of heating for bathing purposes, it is impossible to say, further than that experience demonstrated the fact without being able to point out the cause. Further, there is a vast difference in the quality of heat radiated from different substances. There is nothing really better as a radiating surface than ordinary firebricks. From this material a soft heat is given off, differing in quality from that obtained from iron or any other metallic surface. It would naturally follow from the foregoing statement that the radiant heat given off by steam pipes is vastly inferior to that from radiating surfaces of firebricks, yet, in spite of this fact, there is little likelihood of firebricks coming into extensive use for this purpose, although they may be used in the more expensive establishments. The simplicity, economy and convenience with which steam can be used as a heating medium, and the variety of forms of heating surfaces which can be adapted to any and all possible conditions, makes this the accepted method of heating the sweat rooms of Turkish bath establishments.

There is another reason why radiant heat is better than hot air for heating the sweat rooms. In the bath the object is to keep the head cool and the feet warm. To accomplish this bathers sometimes place a cold, wet towel on the head and sit with their feet in a vessel of hot water.

When the bath is heated by direct radiation the heating surface can be so disposed that most of the heat will be at the bottom of the rooms. With such a disposition the feet will be equally as warm, if not warmer, than the head. Where the pipes are concealed back of firebrick walls and under firebrick floors most of the heat can be radiated from the floor and the balance from the side walls. When the bath is heated by hot air, on the other hand, whether from a furnace or by means of indirect steam, the coldest air in the rooms will be found near the floor, so that the head might be hot while the feet are cold. Usually there is a difference of several degrees between the temperature at the floor line and at the head level, and this difference in temperature is in the reverse order from what it should be.

Steam will be found so necessary around a bathing establishment for operating the machinery to run ventilating fans, laundry machinery and for drying clothes that there is but little likelihood of the hot-air method, with all its disadvantages, being used in this country, consequently the hot-air method of heating sweat rooms will not be considered.

Heating Sweat Rooms with Steam Coils.— The simplest method of heating the sweat rooms of a Turkish bath with steam, and the one commonly followed in practice, is to cover the walls and ceilings of the rooms with a grid of steam pipes made up in the form of coils. In arranging the heating surface in the several rooms it is well to bear in mind the fact that the greatest amount of heat is required near the floor line, so that if enough radiation will not be required to cover both the walls and the ceilings the walls should be covered first and whatever excess radiation then remains can be placed under the ceiling. For convenience in regulating the temperature of the rooms the pipe coils should be put up in sections, so that any or all of them can be cut out of service at any time, either with a view of regulating the temperature of the room or for repairs.

In proportioning the heating surface sufficient radiation will have to be supplied to heat the various compartments to their regular temperatures in the severest winter weather. Such an amount of radiation, of course, would be too much in the summer time, or at intermediate periods between the two extremes, so the best way to arrange the coils is to provide a certain amount of radiation for the summer months, which can be kept in service at all times, winter and summer, then have additional sections in reserve which can be put in service as the weather grows cold until, in midwinter, all the sections are in use. By such an arrangement the heating system is made sufficiently elastic to take care of the sweat rooms in any and all kinds of weather without constantly fussing with the valves and without the annoyance experienced when steam has been shut off all or a great part of the system because the rooms are too hot, only to have them go to the other extreme and fall below their regular temperature.

Types of Coils for the Sweat Rooms.—Whatever type of coil is used for heating the sweat rooms, provision for expansion and drainage must be made when building them and putting them in place. If the pipes are screwed together rigid, without room for expansion, they will bend and buckle when the steam is turned on and the tremendous stress under which the coil is placed might burst some of the fittings. Such an accident would not be a small matter by any means, for high-pressure steam is required for heating the rooms, and if high-pressure steam were to escape from a burst steam coil while bathers were in the room some of them would doubtless be badly scalded before they could make their escape.

In order to prevent rattling and snapping sounds within the coils when the steam is turned on, all the pipes should be given a pitch of from ½ inch to 1 inch in 10 feet, sloping continuously from where the steam enters toward the outlet of the coil. If the coils are not properly drained there will not only be a disagreeable rattling and snap-

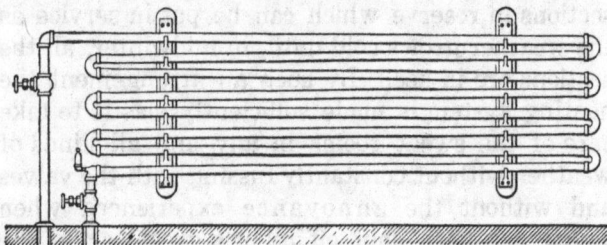

Fig. 20
Continuous Flat Steam Coil

ping sound within, but the coils will not heat as they should and the rooms cannot be maintained at their proper temperature.

A continuous flat coil extensively used as a radiating surface is shown in Fig. 20. In this type of coil, which is also known as a return bend coil, the circulation is direct and positive, steam entering the top pipe and traveling continuously, alternately right and left with the water of condensation, toward the outlet at the bottom. When properly proportioned this is a very efficient form of radiating surface, but it must be properly proportioned or it will prove more or less a failure. Once the steam enters this type of coil it must travel continuously throughout its entire length, and if the coil be extremely long, contains a great number of bends and be exposed in a fairly cold place, the steam will be condensed in the upper loops of the coil, thereby parting with most of its available heat, after which the water of condensation will have to flow through the balance of the coil, where it will part with but little of the heat it contains. It will thus be seen that the lower part of this type of coil under such conditions would be of but little service. Further, owing to the distances the water of condensation would have to travel and the friction of the pipe and bends, the lower part of the coil, particularly if made of 1-inch pipe, would be likely to become overtaxed, or air may be accumulated in the return bends to impede or wholly stop circulation. This form of coil can safely be used for small spaces in the sweat room, but if fairly large coils are to be used it is well to see that they are made of large size pipes, 1½ or 2 inches in diameter, and that the pipes all have a good pitch from where steam enters to the return connection.

Return manifold coils are of more direct circulation than are continuous coils, consequently they are the more efficient type. A return manifold coil is shown in Fig. 21. Steam enters the top end of the upper manifold header and flows almost simultaneously into the several steam loops of the coil. At the end of the steam loops the condensation and vapor flow to the return portion of the coil and pass through the several pipes to the return header, thence through the return pipe to the boiler. It will be noticed that with this type of coil the steam does not have to pass consecutively through all the pipes, but, on the contrary, each steam pipe has a corresponding return loop so that the steam has to pass through only one loop, or twice the length of the coil, before reaching the return header.

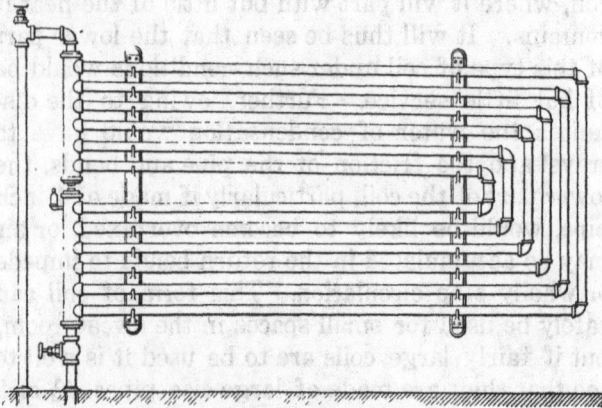

Fig. 21
Return Manifold Steam Coil

Such an arrangement of the pipes insures perfect and immediate drainage of the coil, so that good live steam can fill the pipes at all times, thereby insuring the maximum efficiency. Provision is made for expansion and contraction by having separate headers for the steam and return pipes, which allows considerable spring to the horizontal steam and return pipes of the coil.

Angle wall coils are very convenient in the sweat room for the reason that they can be extended in one piece around two, or even three, sides of the compartment.

An angle wall coil is shown in Fig. 22. This illustration shows a coil intended for only two sides of a room, but by replacing one of the manifolds

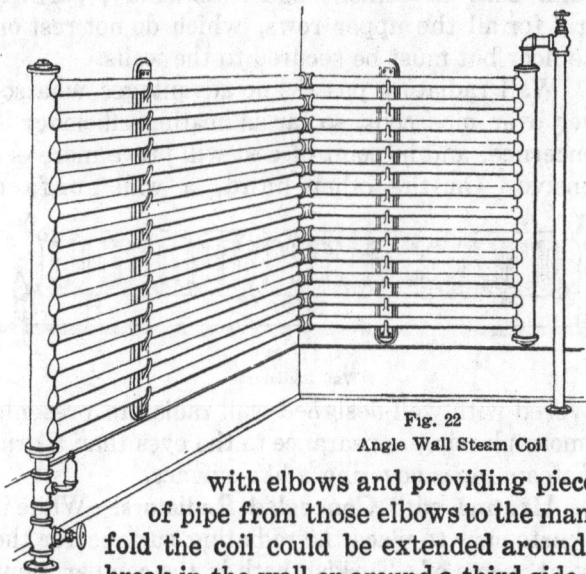

Fig. 22
Angle Wall Steam Coil

with elbows and providing pieces of pipe from those elbows to the manifold the coil could be extended around a break in the wall or around a third side of the room.

In angle wall coils the flow of steam is short and direct, simply from the inlet manifold through one full length of the parallel pipes to the return manifold.

Expansion and contraction of the coil are taken care of by the elbows and pipes, which will give or

67

spring a sufficient amount to compensate for any contraction or expansion of the metal.

Instead of heating the sweat rooms of Turkish baths with pipe coils they may be heated with cast-iron radiators or any other form of steam-heating surface. If cast-iron radiators are to be used wall radiators similar to those shown in Fig. 23 will be found both convenient and satisfactory, particularly for all the upper rows, which do not rest on the floor but must be secured to the walls.

Wall radiators possess no advantages whatsoever over pipe coils, so far as heating efficiency is concerned, and in point of cost will prove more expensive. On the other hand, a wall surface

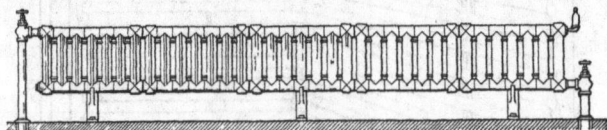

Fig. 23
Wall Radiators

covered with well-designed wall radiators presents a more pleasing appearance to the eyes than a grid of steam pipes covering a like surface.

Heating with Concealed Radiators.—While it is customary to place the radiating surface for the sweat rooms of a Turkish bath in the compartment it is to heat, such a practice is not necessary and is followed more for economic than for other reasons. It cannot be said that wall coils or radiators add anything to the ornamental features of a room, so whenever it is desired, visible radiation may be omitted and concealed radiation installed in its stead. To do so, curtain walls, preferably of firebricks, which may be enameled on the inside surface, are built in front of the steam coils — or, in other words, the coils are set up in a space provided for them back of the firebrick curtain walls and under the floor. Sufficient heating surface is supplied to raise the temperature of the firebricks, which then radiate heat to the bathers.

The surface back of the steam coils must be well insulated and reflectors may be used to throw the heat against the firebrick walls.

Heating a Sweat Room in the Home. — If it were realized how easy it is to fit up a sweat room in the home, no home would be without the benefit of a Turkish bath. As has been pointed out before, no agency but heat is needed for the purpose, water being obtainable in almost every home of to-day. The shower bath is fast becoming an indispensable part of every bath room, and a combined shower bath and sweat room can easily be constructed. Partition off from the main bath room a small square or round enclosure, about three or four feet square, or of the same diameter. Line the walls with tile, marble or porcelain, and set a porcelain or porcelain lined receptor in the floor. Then fit up this small enclosure with sufficient heating surface so that when the steam or hot water is turned on the compartment can be raised in temperature say to 140° Fahrenheit. With a chair placed inside of this enclosure it makes a perfectly satisfactory and efficient sweat room, and the cold shower can be used instead of the plunge for closing the pores after heat enough has been applied. Wall coils around the sides of the enclosure, either concealed or exposed will furnish the heat necessary for the operation.

In case the building is heated with hot air, instead of wall coils an extra large hot-air register with an extra large hot air duct direct from the furnace to the hot room will supply the necessary heat, which can be shut off when the bath is not in use.

In old buildings where no provision has been made for a shower bath or a sweat room, a bath cabinet can be had which will serve the purpose very well. These bath cabinets are more in the nature of Russian baths than Turkish baths, as vapor is used as the heating medium. A valuable feature of these vapor cabinets is the fact that the head is outside the cabinet so that pure cool air is being inhaled, while heat is being applied to the body inside of the cabinet enclosure.

CHAPTER VI

PROPORTIONING THE HEATING SYSTEM

PRINCIPLES OF HEATING

PROPERTIES OF HEAT

TRANSFER of Heat.—When two bodies of different temperatures are brought near to each other a transfer of heat takes place from the hotter to the colder body. This tendency toward maintaining an equilibrium of temperature is universal, and the transfer of heat may take place in any of three different ways—by *conduction,* by *convection* or by *radiation.*

Conduction of Heat.—Conduction of heat is the progressive movement of heat through a substance without perceptible movement of the molecules. If one end of a poker be held in a fire the other end will become heated by conduction. The hot rooms of a Turkish bath lose their heat by conduction, by convection and by radiation. If the walls are not properly insulated a great quantity of heat will flow by conduction from the inner to the outer surface of the wall, where it may radiate to surrounding objects of lower temperatures or be cooled by convection, currents of air carrying off the excess heat.

Convection of Heat.—Convection of heat is the transfer of heat by movement or circulation of the substance to be heated or cooled, which by constantly changing its position brings all parts consecutively into contact with the heating or cooling medium. Water in a vessel placed on a stove is heated by convection, or circulation, the constant movement of the water bringing all particles to the bottom, where the heat is applied. Water in a range boiler is likewise heated by convection, the constant circulation of water to and from the waterback bringing all particles there to be heated, and this heated water gives off part of its heat to that in the boiler when they are brought into contact. In like manner the sweat rooms of Turkish baths warmed by hot air are heated by convection, whether the air from the rooms is circulated through the heater—which, of course, should not be done—or whether fresh air from the outside is forced into the rooms. The air becomes heated by contact with the heating surface in the furnace, and when poured into the sweat rooms parts with some of its heat to the walls, floor and ceiling of the enclosure.

In like manner heat is lost from the sweat rooms of a Turkish bath by currents of cold air blowing against the walls, thereby carrying off the

72

heat that has been transferred by conduction from the interior of the rooms.

Radiation of Heat.—Radiation of heat is the transfer of heat through space from a warm body to one of lower temperature, without the use of intermediate agencies. For example, the earth is warmed by radiation from the sun. Radiant heat does not heat the air through which it passes; it travels direct and in straight lines until intercepted and reflected or absorbed by some other body. The cooler body will reflect or absorb all the heat rays it intercepts, and the sum of the absorption and reflection equals the total of the intercepted rays. When a bather stands naked in the sweat room of a Turkish bath radiant heat from the steam coils passes direct to his body without losing any of its intensity to the air through which it passes.

Heat and Temperature.—In all natural phenomena there are two distinct conditions or properties which can generally be measured or gauged. For instance, there are both pressure and quantity of water, the pressure being measured in pounds per square inch and the quantity in gallons or cubic feet. Electricity is measured in volts and amperes. Sound, likewise, has its intensity and volume, and the condition known as heat has two properties— heat and temperature. The temperature of a substance may be considered the intensity of the energy, and the heat itself as the quantity. The intensity of heat may be measured by means of a thermometer, ordinary temperatures being indicated by the common household thermometer. The quantity of heat, on the contrary, cannot be measured by instruments or by any of the known standards for weight or quantity, such as pounds or gallons, so a new standard had to be devised which would measure the quantity of heat by the effect it would produce. The standard for the measurement of heat is known as a British thermal unit.

British Thermal Unit.—A British thermal unit, usually abbreviated B. T. U., is the quantity of heat required to raise the temperature of one pound of water from 62° to 63° Fahrenheit. In practice it is taken as the quantity of heat required to raise one pound of water 1° Fahrenheit. British thermal units are likewise commonly called heat units.

Heat Emitted by Radiators and Coils.—The quantity of heat transmitted from coils or radiators to air is practically independent of the materials used, so long as they are under similar conditions of internal and external temperature. That is, one square foot of cast-iron surface will give off practically the same quantity of heat as one square foot of wrought-pipe surface, if both the cast-iron radiator and the pipe coil are supplied with heat from steam of the same pressure and are exposed in rooms having equal temperatures. Further, under the same conditions the quantity of heat given off from a hot water radiator will be equal to that given off from an equal size steam radiator. Again, the rate at which heat is emitted from radiating surfaces is not affected by the size of the tubes, or sections, of the radiators, or the diameters of pipe used in making the coils, so long as the pipes, loops or sections are large enough to afford good circulation.

Painting or bronzing radiating surfaces has a marked effect on the quantity of heat emitted, the effect depending on the character of the finished surface. As a rule, it may be stated that painting radiating surfaces with a dull finish or a rough bronzing increases the total quantity of heat given off but decreases the amount of radiant heat emitted. This is important in Turkish bath work, for the reason that in the sweat rooms it is radiant heat, not connected heat that is wanted. However, there seems no practical way of overcoming the difficulty, for it is necessary to coat the pipes and radiating surfaces with something to keep them from rusting, even though it were not desirable for ornamental purposes.

Paints or coatings of any kind having a glossy surface, particularly a glossy white finish, should be avoided, as it reduces not only the radiant heat but the total quantity of heat emitted.

Radiators or coils having but one row of tubes are more effective in still air than those having two or more rows. This fact should be borne in mind when designing the heating sytem for the Turkish bath, and only one-tube coils used unless enough heating surface cannot be installed without doubling up the rows, in which case the double-row coils should be near the floor.

Horizontal coils are more effective than vertical coils in still air, so that all the piping in the sweat room should be run in a horizontal position.

75

In still air, such as obtains in all places where direct radiation is used, low radiators or coils are more effective than tall ones of equal surface. This fact has an important bearing on the heating of Turkish bath rooms, for the reason that in the sweat rooms the coils generally extend from floor to ceiling. The loss in effectiveness is general, radiant heat and convective heat both falling below the normal. The reason for this is that only part of each steam pipe in the coil is so placed as to make available all of the heat it emits. Part of the heat given off as radiation strikes the next lower and higher pipes in the coil and is reflected to the emitting pipe or to the wall back of the coil, and another part of the heat radiates to the wall to which the coil is attached, and from there is reflected back to the pipe which emitted it, or to the one next higher or lower in the coil.

The convected heat suffers in a like degree. The coolest air in the room is near the bottom, and the cooler the air the greater the heat emission of the pipe; consequently, when air at the bottom of the coil becomes heated it flows up toward the ceiling alongside of the steam coil, growing hotter as it progresses, thereby forming a film of hot air in front of the coils, which absorbs but little heat and at the same time prevents cooler air from coming in contact with the pipes.

Emmissive Capacity of Radiators and Coils.—The actual quantity of heat given off by a radiator or coil depends upon the difference in temperature between the steam within the pipe and the air surrounding the coil or radiator.

If, for instance, a portion of a steam pipe containing one square foot of radiating surface were extended through the insulated side wall of a building into still air, so long as the temperature of the air and steam remained constant the steam pipe would give off, hour after hour, an equal amount of heat. If a well-insulated box were next placed over the steam pipe, as shown in Fig. 24, and the steam were maintained at the same temperature, heat would begin to radiate to the walls of the box, and air passing over the surface of the steam pipe would abstract more heat, which would be given off, assume, at the rate of 300 B. T. U. per hour. At that rate of emission the box and the air contained therein would soon begin to rise in temperature, and, in proportion as the temperature of the box and air increased, the emissive capacity of the steam pipe would decrease correspondingly until a point was reached where just enough heat would be given off to supply that lost by radiation and convection from the outside of the box. When this point was reached, it would be found that instead of 300 B. T. U., the heat emission of the pipe had dropped to perhaps 30 B. T. U. per hour.

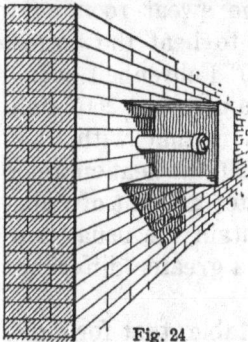

Fig. 24

Steam Pipe Through Wall

The actual amount of heat given off per square foot of surface has been found by experiment to be constant for each degree of difference between the

77

temperature of the air surrounding the coils and the temperature of the steam within the pipes, and from this experimental data rules have been devised for finding the amount of radiating surface required to heat given spaces.

The rate of emission and the number of heat units (B. T. U.) emitted per hour per square foot of radiating surface for various degrees of difference between steam and air, in horizontal tube radiation, in still air, may be found in Table I.

To better understand the value of the table apply it to the heating of the sweat room of a Turkish bath. For example, to heat the hottest room in a Turkish bath to 240° Fahrenheit would require steam at a temperature of about 260° Fahrenheit. The reason for this is that with a less difference in temperature than 20° the heat emission drops so low that a much greater amount of radiation would be required to maintain the usual temperature of the rooms than if a greater difference in temperature be maintained.

Note, for instance, in the table, that for a difference of 10° temperature only about one-half heat unit is given off per hour for each degree difference in temperature between the steam and the air of the room, while with a difference of temperature of 20°, 1.11 heat units are given off from the same amount of radiating surface in an equal period of time for each degree difference in temperature.

It may be assumed, then, that the steam used for heating Turkish bath rooms will have a temperature at least 20° higher than that to which the

TABLE I.—HEAT EMISSION FROM HORIZONTAL TUBE RADIATION

	Heat Emitted per Square Foot per Degree Difference Between Temperature of Air and Steam				Total Heat Emitted in B. T. U. per Square Foot per Hour for Degrees Difference Shown in Column 1			
Column 1	Column 2	Column 3	Column 4	Column 5	Column 6	Column 7	Column 8	Column 9
Difference in Temperature Between Air and Steam in Degrees, Fahr.	Six-Inch Pipe Coil 40 Inches High Massed Pipes	Four-Inch Pipe Coil 40 Inches High Single Row Pipes	Two-Inch Pipe Coil 24 Inches High Massed Pipes	One-Inch Pipe Coil 12 Inches High Single Row Pipes	Six-Inches Pipe Coil 40 Inches High Massed Pipes	Four-Inch Pipe Coil 40 Inches High Single Row Pipes	Two-Inch Pipe Coil 24 Inches High Massed Pipes	One-Inch Pipe Coil 12 Inches High Single Row Pipes
10	0.55	0.62	0.66	0.85	5.50	6.2	6.6	8.5
20	1.11	1.25	1.32	1.72	20.2	24.9	26.4	34.4
30	1.18	1.34	1.42	1.84	35.	39.7	42.7	55.2
40	1.24	1.40	1.48	1.92	49.6	56.2	59.	77.
50	1.29	1.46	1.54	2.01	64.5	73.0	77.	100.
60	1.33	1.50	1.58	2.06	79.8	90.	95.	124.
70	1.36	1.54	1.63	2.12	95.2	108.	113.	148.
80	1.40	1.58	1.67	2.18	112.	127.	133.	173.
90	1.43	1.63	1.72	2.24	128.	147.	153.	199.
100	1.47	1.66	1.76	2.28	147.	167.	175.	228.
110	1.51	1.71	1.80	2.34	166.	188.	198.	257.
120	1.54	1.74	1.84	2.39	184.	208.	219.	287.
130	1.57	1.78	1.88	2.44	203.	230.	242.	318.
140	1.61	1.81	1.91	2.48	223.	252.	266.	346.
150	1.64	1.84	1.94	2.53	244.	276.	291.	378.
160	1.66	1.87	1.97	2.57	265.	300.	316.	410.
170	1.69	1.91	2.02	2.62	286.	324.	341.	443.
180	1.72	1.94	2.05	2.65	307.	348.	367.	475.
190	1.75	1.98	2.09	2.71	330.	375.	393.	512.
200	1.78	2.01	2.12	2.76	356.	393.	415.	552.
225	1.87	2.12	2.24	2.91	420.	477.	500.	650.
250	1.97	2.23	2.35	3.06	493.	557.	587.	762.
275	2.07	2.34	2.47	3.22	563.	637.	670.	872.
300	2.17	2.45	2.58	3.37	654.	742.	780.	1,020.
325	2.27	2.55	2.70	3.50	740.	840.	882.	1,160.
350	2.37	2.67	2.82	3.66	835.	945.	995.	1,295.

¹ After Carpenters' "Heating and Ventilating of Buildings"
² Totals divided by 1,000 give approximate weight of steam condensed per hour.

hottest room will have to be heated, and a greater temperature of steam than 20° difference will be found even better.

For the sake of explaining the table, then, assume a temperature of the air of 140° Fahrenheit and a temperature of steam of 260° for one of the hot rooms in the bath, which is to be maintained to a temperature of 140° Fahrenheit.

The temperature of the steam is then 260°, that of the air in contact is 140°, and the difference between the two temperatures would be 260° minus 140°, or 120° Fahrenheit. Glancing down column 1 of the table to the difference of temperature rated as 120° and following the horizontal line to column 3, it will be found that 1.74 heat units are given off per hour from each square foot of radiating surface for each degree difference in temperature between the steam and the air. To find the total amount of heat given off per hour from one square foot of radiating surface with that many degrees difference in temperature all that is necessary is to multiply 120, the difference in temperature, by 1.74, the rate of emission, which gives 208 B. T. U. per hour. Instead of multiplying, the same information will be obtained by following the horizontal line to column 7, which gives the total amount of heat emitted per hour for one square foot of this size of coil at a difference of temperature of 120°.

It will be noticed in the foregoing explanation that the coil is only 40 inches in height and gives off 1.74 heat units per hour, while the 12-inch single row radiation in column 5 of this table gives off 2.39 heat units per hour, or .65 heat unit more than an equal radiating surface 40 inches in height. The same law will hold good throughout and the higher the coils extend the less heat will be given off per hour from one square foot of surface. If, therefore, the coils to be installed will extend from the floor to the ceiling, or a good portion of the way there, allowance would have to be made for that fact, and a less rating given the surface, or the coils will be overrated. Perhaps in the example used in explanation an allowance of 1.5 heat units per hour per square foot for each degree difference would have been more nearly right. Unfortunately, there are no experimental data available showing the actual difference for very high coils.

The size of radiator or coil in square feet required to supply a certain number of heat units per hour, when the temperature between the steam and air is known, can be found by means of the following rule:

RULE.—To find the number of square feet of radiation required to supply a certain number of heat units per hour, divide the total number of heat units to be supplied by the total number of heat units given off per hour from one square foot of surface at the stated difference of temperature.

EXPRESSED AS A FORMULA:

$$\frac{h}{c} = r$$

In which h = heat units to be supplied; c = total heat units per hour given off by one square foot of

TABLE II—PROPERTIES OF SATURATED STEAM

Absolute Pressure	Temperature Degrees Fahr.	Heat Units Above 32° Fahr. Contained in 1 Pound of Steam			Weight of 1 Cubic Foot in Pounds	Volume of 1 Pound in Cubic Feet
		In Water	Latent Heat	Total Heat		
14.7	212 0	180.9	965 7	1146.6	.0379	26.37
15.	213 1	181.6	965 3	1146 9	.0387	25.85
16.	216.3	184.9	963 0	1147.9	.0411	24.33
17.	219.5	188.1	960.8	1148 9	.0435	22.98
18.	222.4	191.1	958 7	1149 8	.0459	21.78
19.	225.3	193.9	956.7	1150.6	.0483	20.70
20.	228.0	196.7	954 8	1151 5	.0507	19 73
21.	230.6	199 3	953 0	1152.3	.0531	18 84
22.	233.1	201 8	951 2	1153.0	.0554	18 04
23.	235.5	204 3	949 5	1153 8	.0578	17.30
24.	237 8	206 6	947 9	1154 5	.0602	16.62
25.	240.1	208 9	946 3	1155.2	.0625	16.00
26.	242 2	211 1	944 7	1155 8	.0649	15.42
27.	244.3	213 2	943 3	1156.5	.0672	14.88
28.	246 4	215 3	941 8	1157 1	.0695	14.38
29.	248 4	217 3	940 4	1157 7	.0719	13.91
30.	250 3	219.3	939 0	1158 3	.0742	13.48
31.	252 2	221 2	937 7	1158 9	.0765	13.07
32.	254.0	223.0	936.4	1159 4	.0788	12.68
33.	255.8	224.8	935.1	1159.9	.0812	12.32
34.	257.5	226 6	933 9	1160 5	.0835	11 98
35.	259.2	228 3	932 7	1161 0	.0858	11.66
36.	260.9	230.0	931 5	1161 5	.0881	11.36
37.	262.5	231 6	930 4	1162 0	.0904	11.07
38.	264.1	233 3	929 2	1162 5	.0927	10 79
39.	265.6	234 8	928 1	1162 9	.0949	10 53
40.	267.2	236 4	927 0	1163 4	.0972	10.28
41.	268 7	237 9	926.0	1163 9	.0995	10.05
42.	270 1	239.4	924 9	1164.3	.1018	9 83
43.	271.6	240 8	923.9	1164.7	.1041	9 61
44.	273 0	242 3	922 9	1165 2	.1063	9.40
45.	274 3	243 7	921 9	1165 6	.1086	9.21
46.	275 7	245 1	920 9	1166 0	.1109	9.02
47.	277 0	246.4	920 0	1166 4	.1131	8.84
48.	278 3	247 7	919.1	1166 8	.1154	8.67
49.	279 6	249 1	918 1	1167 2	.1177	8.50
50.	280 9	250 3	917 3	1167.6	.1199	8.34
51.	282 2	251.6	916 4	1168 0	.1222	8.19
52.	283 4	252.9	915 5	1168 4	.1244	8.04
53.	284 6	254 1	914 6	1168 7	.1267	7.89
54.	285 8	255 3	913 8	1169 1	.1289	7.76
55.	287 0	256 5	912 9	1169 4	.1312	7.62
56.	288 1	257 7	912 1	1169 8	.1334	7.50
57.	289 3	258 9	911 3	1170 2	.1357	7.37
58.	290 4	260 0	910 5	1170 5	.1379	7.25
59.	291 5	261 1	909 7	1170 8	.1401	7.14
60.	292 6	262 3	908 9	1171 2	.1424	7.02
61.	293 7	263 3	908 2	1171 5	.1446	6.92
62.	294 7	264 4	907 4	1171 8	.1468	6.81
63.	295 8	265 5	906 6	1172.1	.1491	6.71
64.	296 8	266 6	905 9	1172 5	.1513	6.61
65.	297 8	267 6	905 2	1172 8	.1535	6.52
66.	298.8	268 7	904 4	1173.1	.1557	6.42
67.	299 8	269 7	903 7	1173 4	.1579	6.33
68.	300 8	270 7	903 0	1173 7	.1602	6.24
69.	301 8	271 7	902 3	1174 0	.1624	6.16
70.	302 8	272.7	901.6	1174 3	.1646	6.08
71.	303.7	273.6	901.0	1174.6	.1668	6.00
72.	304 7	274 6	900.3	1174 9	.1690	5.92
73.	305 6	275 6	899.6	1175.2	.1712	5.84

No.							No.						
74	306.5	276.5	1175.4	898.9	.1734	5.77	115	337.9	308.6	876.4	1185.0	2627	3.81
75	307.4	277.4	1175.7	898.3	.1756	5.69	116	338.5	309.3	875.9	1185.2	2649	3.78
76	308.3	278.4	1176.0	897.6	.1778	5.62	117	339.2	310.0	875.4	1185.4	2670	3.75
77	309.2	279.3	1176.3	897.0	.1800	5.56	118	339.8	310.6	875.0	1185.6	2692	3.72
78	310.1	280.2	1176.5	896.3	.1822	5.49	119	340.4	311.3	874.5	1185.8	2713	3.69
79	311.0	281.1	1176.8	895.7	.1844	5.42	120	341.1	311.9	874.1	1186.1	2735	3.66
80	311.9	282.0	1177.1	895.1	.1866	5.36	121	341.7	312.5	873.6	1186.3	2757	3.63
81	312.7	282.8	1177.3	894.5	.1888	5.30	122	342.4	313.1	873.2	1186.5	2778	3.60
82	313.6	283.7	1177.6	893.9	.1910	5.24	123	343.0	313.8	872.7	1186.7	2799	3.57
83	314.4	284.6	1177.8	893.3	.1932	5.18	124	343.5	314.4	872.3	1186.9	2821	3.55
84	315.3	285.4	1178.1	892.7	.1954	5.12	125	344.1	315.1	871.8	1187.1	2842	3.52
85	316.1	286.2	1178.3	892.1	.1976	5.06	126	344.7	315.7	871.4	1187.3	2864	3.49
86	316.9	287.1	1178.6	891.5	.1998	5.01	127	345.3	316.3	871.0	1187.4	2885	3.47
87	317.7	287.9	1178.8	890.9	.2020	4.95	128	345.9	316.9	870.5	1187.6	2907	3.44
88	318.5	288.8	1179.1	890.3	.2042	4.90	129	346.5	317.5	870.1	1187.8	2928	3.42
89	319.3	289.6	1179.3	889.7	.2063	4.85	130	347.1	318.1	869.7	1188.0	2950	3.39
90	320.1	290.4	1179.6	889.2	.2085	4.80	131	347.7	318.7	869.3	1188.2	2971	3.37
91	320.9	291.2	1179.8	888.6	.2107	4.75	132	348.3	319.3	868.9	1188.5	2992	3.34
92	321.7	291.9	1180.0	888.1	.2129	4.70	133	348.9	319.9	868.4	1188.7	3014	3.32
93	322.4	292.8	1180.3	887.5	.2151	4.65	134	349.4	320.5	868.0	1188.9	3035	3.30
94	323.2	293.5	1180.5	887.0	.2173	4.60	135	350.0	321.1	867.6	1189.1	3057	3.27
95	323.9	294.3	1180.7	886.4	.2194	4.56	136	350.6	321.7	867.2	1189.2	3078	3.25
96	324.7	295.1	1181.0	885.9	.2216	4.51	137	351.1	322.3	866.8	1189.4	3099	3.23
97	325.4	295.8	1181.2	885.4	.2238	4.47	138	351.7	322.8	866.4	1189.6	3121	3.20
98	326.2	296.6	1181.4	884.8	.2260	4.43	139	352.3	323.4	866.0	1189.7	3142	3.18
99	326.9	297.3	1181.6	884.3	.2281	4.38	140	352.8	324.0	865.6	1189.9	3163	3.16
100	327.6	298.1	1181.9	883.8	.2303	4.34	141	353.4	324.6	865.1	1190.1	3185	3.14
101	328.3	298.8	1182.1	883.3	.2325	4.30	142	353.9	325.1	864.8	1190.2	3206	3.12
102	329.1	299.6	1182.3	882.7	.2346	4.26	143	354.5	325.7	864.4	1190.4	3227	3.10
103	329.8	300.3	1182.5	882.2	.2368	4.22	144	355.0	326.2	864.0	1190.6	3249	3.08
104	330.5	301.0	1182.7	881.7	.2390	4.19	145	355.6	326.8	863.6	1190.7	3270	3.06
105	331.2	301.7	1182.9	881.2	.2411	4.15	146	356.1	327.4	863.2	1190.9	3291	3.04
106	331.9	302.4	1183.1	880.7	.2433	4.11	147	356.6	327.9	862.8	1191.0	3313	3.02
107	332.6	303.2	1183.4	880.2	.2455	4.07	148	357.2	328.5	862.4	1191.2	3334	3.00
108	333.2	303.9	1183.6	879.7	.2476	4.04	149	357.7	329.0	862.0	1191.4	3355	2.98
109	333.9	304.6	1183.8	879.2	.2498	4.00	150	358.2	329.6	861.6	1191.6	3376	2.96
110	334.6	305.3	1184.0	878.7	.2519	3.97	160	363.3	334.9	857.9	1192.8	3589	2.79
111	335.3	305.9	1184.2	878.3	.2541	3.94	170	368.2	339.9	854.4	1194.3	3801	2.63
112	335.9	306.6	1184.4	877.8	.2563	3.90	180	372.9	344.7	851.0	1195.5	4012	2.49
113	336.6	307.3	1184.6	877.3	.2584	3.87	190	377.4	349.3	847.7	1197.0	4223	2.37
114	337.2	308.0	1184.8	876.8	.2606	3.84	200	381.6	353.7	844.6	1198.3	4433	2.26

surface at temperature stated; r = square feet of radiating surface required.

EXAMPLE.—How many square feet of radiating surface, 40 inches high, single columns, will be required to supply 900,000 heat units per hour, with a temperature difference of 160° between the steam and air?

SOLUTION.—Heat units to be supplied, 900,-000 = h; difference in temperature between air and steam, 160° Fahrenheit; total heat units given off per hour by one square foot of surface at 160° difference (see column 7 in table I), 300 = c.

Substituting those values in the formula: h − c = 900,000 ÷ 300 = 3,000 square feet surface. (Answer).

Pressure and Temperature of Steam.—In order to calculate the size of coils for heating purposes and the kind of engines for operating the machinery and other apparatus about the mechanical installation for a Turkish bath establishment, the pressure and corresponding temperature of the steam to be carried by the boiler must be determined beforehand.   There is a certain intimate relation between the temperature of steam and its pressure, which is absolute.   Pressure cannot be increased without also increasing the temperature of the steam and the boiling point of the water, nor can the temperature of the steam and boiling point of the water be increased without increasing the pressure.   The temperature and pressure of boiling water and the temperature and pressure of steam in contact are always equal.   This relation between pressure and temperature of steam is important, and convenient data, when designing the bath, for by referring to a table of properties of saturated steam the right temperatures and pressure can be bound at a glance.

For instance, if a pressure of 60 pounds per square inch is necessary for operating the various engines for pumps, blowers, laundry machinery, etc., a glance at the table shows that the temperature corresponding to that pressure is 307° Fahrenheit, which is sufficiently high for all purposes.

The temperature of steam at different pressures can be found in Table II. In using this table either add 14.7 to the reading of the gauge, to get absolute pressure, or to get gauge pressure deduct 14.7 from the absolute pressures given in the table. Usually 15 pounds added to gauge pressure or subtracted from the absolute pressures will be sufficiently accurate for all ordinary purposes.

Comparative Efficiency of Different Radiators.—It might be well to emphasize the fact here that all radiators are not of equal value. As has already been pointed out, wrought pipe is more efficient per square foot of heating surface than cast-iron radiators, and cast-iron radiators differ according to their design, so that in still air, which would mean all direct radiation, some designs or types of cast-iron radiators are far more effective than others, and what would be 150 square feet of heating surface in one, might have the heating capacity of only 100 square feet of surface, although as a matter of fact it actually has 150 square feet of surface. It will be seen, then, that the difference between 150 square feet and 100 square feet in such a case instead of representing heat value would represent only weight and pig iron, and should not be classed as radiation.

This difference in cast-iron radiation depends upon two things; height of the radiators, and number of columns or loops to the radiator sections. A 20-inch radiator will emit more heat per square foot of surface than a 23-inch radiator; a 23-inch radiator more than a 26; a 26 more than a 32; a 32 more than a 38; and a 38 more than a square foot of surface on a radiator 45 inches in height. It will be seen, therefor, that in selecting the radiators it is well to keep the height of loop in mind, and see that only moderately high loop radiators are used without making due allowance for the less heating value of the higher radiators.

Again, the greater the number of columns the radiator has per section, the less its heat value per square foot of surface. Of all radiation, a one column type is the most effective. Next to the one loop comes the two-loop, after the two-loop radiation comes the three loop, then the four loop, and lastly the five loop. Keep in mind, then, that the one column low radiator is the most effective design made.

CHAPTER VII

LOSS OF HEAT FROM THE BUILDING

AUSES of Heat Loss.—Heat is lost from buildings by leakage around windows, doors and other openings, by conduction through the walls of buildings, and by passing out through the ventilation registers with the vitiated air. Heat losses from leakage can be neglected in calculating the heat to be supplied for Turkish bath buildings when the walls have been properly insulated, as they should be, according to methods previously described, and heat losses through the ventilation registers can be disregarded in calculating the amount of radiation required to heat the rooms, for such losses will be made up by the heat supplied in heating the air used for ventilation purposes. That leaves as the only heat loss to be ascertained, the heat which passes off from the exposed walls of the building. This heat loss may be divided into two parts—that lost by radiation from the wall to surrounding objects and that lost by convection, or carried off by the cold air or winds which come in contact with the outer walls. Of the two losses, that by convection is by far the more serious, as it is the greater and the more uncertain, depending upon exposure and wind currents, and will average, perhaps, 75 per cent. of the total heat lost from the walls.

The actual heat lost from the walls of a Turkish bath building depends on the character and thickness of the walls or surface, the materials of which they are built, the difference between the temperature of the air inside and that outside and the exposure. In many respects there is a great similarity between the heat transmitted by radiators and coils and the heat lost from buildings. The chief point of similarity lies in the fact that the heat given off in both cases depends upon the difference in temperature between the gases or fluids on opposite sides of the dividing wall or surface. There is this marked difference between them, however; in the case of radiators and coils the heating surface is a good conductor which practically interposes no barrier against the transmission of heat. The walls of Turkish bath buildings, on the other hand, are made to retard as much as possible the transmission of heat, so that under a difference of temperature between steam and air which would cause a heat emission of 300 heat units per hour per square foot of surface, well insulated walls would give off only about 10 per cent. of that amount from the same area of surface.

The amount of heat which is lost through various substances and from walls and partitions such as are generally used in building construction has been determined by the German government and the results translated to American equivalents by Mr. Alfred R. Wolff, M. E., while the results of experiments and tests made by Péclet have been reduced to American equivalents by Professor Carpenter. Unfortunately, however, many of the results are indefinite for the present purpose, as not pointing out the exact kind of construction used, while not any of the tests were made to determine the heat losses from walls such as would be used for Turkish bath rooms, and no data are available giving the heat loss from concrete walls, a form of construction now in common use and one which is a splendid heat retardant. From the American equivalents translated and computed by Wolff and Carpenter the approximate heat losses from walls and materials of various kinds suitable for Turkish bath construction have been compiled in tabular form and are here given as a guide in calculating the heat lost from like constructions, until such time as experimental data will make available the exact values to use.

Loss of Heat from Windows.—The windows of a building are one of the most extravagant sources of heat loss, unless they are carefully fitted so air cannot pass in or out around the window frames or leak through between the glass panes and the sash. Even then the heat loss will be considerable from conduction, unless more than one thickness or layer of glass is used, with air spaces between. When this is done—that is, when two or more layers of glass are used—window surfaces of Turkish bath buildings can be made no better conductors of heat than ordinary 4 to 8-inch brick walls. The rate of heat loss from window surfaces, with different number of thicknesses of glass, can be found in Table III. This table gives the loss in heat units per hour per square foot of exposed surface for each degree of difference in temperature between the air of the room and that outside.

TABLE III.—HEAT LOSS FROM WINDOWS

KIND OF WINDOW	Loss in Heat Units per Hour per Square Foot of Surface for Each Degree Difference Between the Air of the Room and the Air Outside.
Single Window.	1.09
Double Sash, with air space between ...	.52
Three layers of glass, with two air spaces between	.41
Four layers of glass, with three air spaces between	.33
Five layers of glass, with four air spaces between	.25

The values given in the table, it must be remembered, show the rate of heat loss, not the total loss. The total loss from a window surface can be found by the following rule:

RULE.—To find the total heat loss from a window surface multiply the difference in temperature between the inner and the outer air by the rate of loss given in the table for the number of layers of glass and multiply the product thus obtained by the total number of square feet of surface in the window. The final quotient will be the total number of heat units lost by the window surface per hour.

EXPRESSED AS A FORMULA:

$$t \; l \; s - h$$

when t is the difference in temperature between the inner and outer air, l the rate of loss for kind of window given in table, s the square feet of surface in the window, h the total heat units lost.

EXAMPLE.—What will be the total heat loss from a window 4 feet wide and 8 feet high, containing three layers of glass with air spaces between, when the difference between the air of the room and the outer air equals 180° Fahrenheit?

SOLUTION.—t (temperature difference between inner and outer air) = 180; l (rate of loss for three-layer glass) = .41; s (square feet of window surface) = 4 × 8 = 32, substituting these values in the formula:

188.× .41 × 32 = 2361.6 B. T. U. (Answer).

Loss of Heat from Skylights.—The loss of heat from skylights varies but little in quantity from the loss of heat from windows, the difference being a trifle greater for skylights, due, probably, to the fact that heated air can get away quicker from a sloping surface than from a perpendicular one, thereby bringing a greater amount of air into contact with the skylight, with consequently greater cooling effect. The loss of heat from sky-lights of various layers with air spaces between can be found in Table IV. This table, like the preceding one, gives the rate of loss in heat units per hour per square foot for each degree difference in temperature between the air inside and that outside.

TABLE IV.—HEAT LOSS FROM SKYLIGHTS

CHARACTER OF SKYLIGHT	Number of Heat Units Given off per Hour from each Sq. Foot of Surface for Each Degree Difference Between the Temperature of Air Outside and Inside.
Single layer..	1.118
Two layers, with air space between..........	.62
Three layers, with two air spaces between	.53
Four layers, with three air spaces between	.44
Five layers, with four air spaces between	.36

To find the total amount of heat loss use the rule, or formula, given, simply substituting values given in the table of heat loss from skylights for those of windows.

Exposure of skylights to the direct rays of the sun would have a beneficial effect in keeping the heat loss down during the daytime of bright weather, but, as the interior compartments of the building must be maintained at the regular temtemperatures night and day during all kinds of weather, the beneficial effect of the sun in this respect cannot be considered.

In the two foregoing tables the heat losses have been given for windows and skylights having five thicknesses of glass. It is doubtful, however, if, under ordinary conditions, more than three thicknesses will be required except in the most northern parts of the United States, including

Alaska, and Canada. In such climates, where the mercury often falls to 20° below zero and stays there for weeks at a time, economy might dictate more than three layers of glass. Further, if in other parts of the United States a Turkish bath house is to be built in an exposed place, with a prevailing strong north or western wind blowing against the exposed walls of the hot rooms in which windows are located, more than three thicknesses of window glass might be advisable.

Loss of Heat from Walls.—The loss of heat from walls of various kinds and thicknesses can be found in Table V. In this table the values given are per hour per square foot of surface for each degree difference between the temperature of the air inside and that outside.

From the various thicknesses of walls and the number of air spaces listed in the table it ought to be possible to find some one construction sufficiently similar to the one being planned to give an approximate heat loss to use as a coefficient. If more air spaces are provided than shown in the table a deduction of from .07 to .08 heat unit will be about right for each additional air space. Air spaces were not listed in the table for walls of greater thickness than 24 inches, for the reasons that in present-day construction it is not likely that anything but foundation walls will exceed that thickness, and if the Turkish bath apartments are to be located below the street level, where the walls are over 28 inches thick, the heat loss will be so small that an air space would hardly be necessary.

TABLE V.—HEAT LOSSES FROM WALLS

CHARACTER OF WALLS	Thickness of Walls in Inches	Heat Units Lost per Hour from Each Sq Foot of Surface for Each Degree Difference Between the Temperature of Air Inside and Outside.
Ordinary wall of frame building	4	.24
Solid stone, brick or concrete	4	.68
Solid stone, brick or concrete	8	.46
Stone, brick or concrete, with air space	8	.38
Solid brick, stone or concrete .	12	.32
Stone, brick or concrete, with air space	12	.25
Solid stone, brick or concrete	16	.26
Stone, brick or concrete, with air space .	16	.20
Solid stone, brick or concrete	20	.23
Stone, brick or concrete, etc., with air space	20	.16
Stone, brick or concrete, with two air spaces	20	.09
Solid stone, brick or concrete	24	.20
Stone, brick or concrete, with air space	24	.13
Stone, brick or concrete, with two air spaces	24	.06
Solid stone, brick or concrete ..	28	.174
Solid stone, brick or concrete	32	.15
Solid stone, brick or concrete . .	36	.129
Solid stone, brick or concrete . .	40	.115

An important consideration to bear in mind is that all of the values for heat losses given in the foregoing tables are for still air in sheltered places. If the building is to have a northern or western exposure and be subjected to strong winds an allowance of 10 per cent. should be added to the values of heat losses for the walls, windows or skylights so exposed.

Loss of Heat from Floors and Ceilings.— The loss of heat from ceilings of certain construction will be found greater than the loss of heat from floors of similar construction. This difference in heat loss is due to two causes. In the first place, as heat tends to escape upward, any which is conducted downward through the floors has less chance of escaping by convection than has heat which has passed upward to the floor above through the ceiling. In consequence of this fact the rate of transmission or heat losses upward through the ceiling is greater than that downward through the floor. Further, the temperature of the air at the top of a Turkish hot room will be found to be from $10°$ to $15°$ Fahrenheit higher than the air near the floor, the exact difference depending greatly upon the height of ceiling, and as there is a less difference of temperature between the air at the floor line and that of the ceiling below than between the air at the ceiling line and that of the floor above there would naturally be a less loss of heat per square foot, even though the rate of loss were the same for both floors and ceilings.

The loss of heat through floors and ceilings of wooden and fireproof construction can be found in Table VI.

The values given in this table are rather vague and indefinite, as not pointing out the character of fireproof construction referred to. However, the coefficients will serve as a guide to the designer, who must use them with judgment. The values

95

TABLE VI.
HEAT LOSS FROM FLOORS AND CEILINGS

CHARACTER OF SURFACE	Heat Units Lost per Hour from Each Sq. Foot of Surface for Each Degree Difference Between the Temperature of Air on Opposite Sides.
Wooden floor, beams planked over; for floor surface	.083
Wooden floor, beams celled for ceiling surface	.104
Fireproof construction, floored over as flooring	.124
Fireproof construction, floored over as ceiling . ... b	.145

given are for still air, and as the floors and ceilings cannot possibly be exposed to wind currents under ordinary conditions no allowance need be made as in the case of the preceding tables.

CALCULATING HEATING SURFACE

Amount of Radiation Required.—Having the rate of heat emission from radiators and coils and the loss of heat from the walls of buildings, the next step is to determine the amount of heating surface which will be required to supply that heat loss. This, perhaps, can be explained best by a concrete example, showing how to attack the problem. Assume, then, that a Turkish-bath establishment is to have but one hot room, which will be heated to a temperature of 180° Fahrenheit. This room will be 30 feet long, 20 feet wide and 15 feet high. One side and one end, each having 16-

inch solid brick walls, are exposed in northerly and westerly directions and will be swept by strong winds. A window 6 by 8 feet, with three layers of glass, is located in the side and one in the end wall. The other two walls are of 8-inch brick and adjoin compartments which will be heated to 110° Fahrenheit. The floor and ceiling are of fireproof construction and are exposed in rooms having temperatures of 70° Fahrenheit. The problem, then, is to find the amount of radiating surface required to keep the rooms at a temperature of 180° in zero weather, with steam supplied under a pressure of 60 pounds per square inch.

Solving the problem, the following square feet of surface will be found: 654 square feet of 16-inch outside solid wall; 96 square feet of 3-layer window surface; 750 square feet of 8-inch solid interior wall surface; 600 square feet of fireproof ceiling surface; 600 square feet of fireproof floor surface, and the following temperature differences will be ascertained:

Difference between interior air and outside air, 180° Fahrenheit; difference between air of sweat room and other compartments, 180° − 110° = 70°; difference between air of sweat room and that on other sides of floor and ceilings, 180° − 70° = 110°.

The rate of transmission through 16-inch solid walls is .26 heat unit per hour per square foot for each degree difference in temperature between the air inside and outside. This difference of temperature is 180°, therefore each square foot of exposed wall surface will lose .26 × 180, or 46.80 heat units per hour in still air. But owing to the exposures of the walls 10 per cent. must be added to the heat loss in still air, so that the rate of heat loss from the outside walls will equal 46.80 × 1.10, or 51.48 heat units per hour from each square foot of surface. As there are 654 square feet of exposed wall surface in the building the total heat loss from the outside walls will be 654 × 51.48, or 33,667 heat units per hour.

The rate of heat loss through three layers of glass is .41 heat unit per hour from each square foot of surface for each degree difference between the temperature of the air inside and outside. At a temperature difference of 180° the heat loss per square foot per hour would be 180 × 41, or 73.80 heat units.

But the windows are exposed as are the walls, so that 10 per cent. must be added, bringing the heat loss up to 73.80 × 1.10, or 81.18 heat units per hour from each square foot of surface. There are 96 square feet of glass surface exposed, so the total heat loss from windows would equal 96 × 81.18, or 7,793 heat units per hour.

Heat loss through 8-inch solid brick walls is at the rate of .46 heat unit per hour per square foot of surface for each degree difference between the air on the opposite sides of the wall. In this case, the difference in temperature is 180° − 110°, or 70° Fahrenheit. At a difference of 70° of temperature and a heat loss of .46 heat unit the total heat loss per hour from one square foot of surface would be 70 × .46, or 32.20 heat units. This wall is protected so that no allowance need be made for exposures or winds, consequently the total heat loss from the entire wall would be 750 × 32.20, or 24,250 heat units per hour.

There are 600 square feet of fireproof floor surface, losing heat at the rate of .124 B. T. U. per hour per square foot of surface for each degree difference in temperature, and there is a difference in temperature of 180° − 70°, or 110°. The heat loss per hour, then, would be 110° × .124, or 13.64 heat units per square foot of surface, and 13.64 × 600, the total number of square feet of surface, gives 8,184 heat units as the total loss through the flooring.

The ceiling has the same number of square feet of surface as the floor, and the difference in temperature is the same, the rate of loss only being different. If, then, the rate of transmission for ceilings, .145, be substituted for .124 in the preceding explanation or solution for floor surface, it will give 110 × .145 × 600, or 9,570 heat units as the total loss from the entire ceiling surface.

Combining the heat losses from the several sources then gives:

	Heat Units
Heat loss from 650 square feet of 16-inch wall surface	33,667
Heat loss from 96 square feet of three-layer window surface	7,793
Heat loss from 750 square feet of 8-inch brick wall surface	24,250
Heat loss from 600 square feet fireproof floor surface...	8,184
Heat loss from 600 square feet fireproof ceiling surface.	9,570
Total heat loss from the compartment	83,464

It might be well to point out here that as 24,250 heat units per hour pass from the hot room to an adjoining compartment an equal number of heat units may be deducted from the number required to heat that compartment.

The total number of heat units lost, which equal the number of heat units required to heat the sweat room (that required for ventilation excepted), have now been determined, and it only remains to find out how many square feet of heating surface will be required to supply that amount of heat. Referring back to the rule for finding the number of square feet of heating surface required to supply a certain number of heat units per hour, it will be found that all that is necessary is to divide the total heat loss per hour by the total number of heat units given off per hour from one square foot of heating surface at the difference of temperature between the air of the room and the steam in the pipes.

In the present explanation steam is to be supplied at a pressure of 60 pounds per square inch, which corresponds to a temperature of 307° Fahrenheit, and the difference between that temperature and the temperature of the room, which will be 180°, is equal to 307° — 180°, or 127° Fahrenheit. At this difference of temperature one square foot of coils 40 inches high will emit 230 heat units per hour; then, 83,464 ÷ 230 = 362 square feet of radiation.

If the coils are higher than 40 inches an allowance would have to be made for that fact and a lower heat emission assumed, which would increase the total heating surface required.

In the foregoing explanation no allowance has been made for the heat required for the air in the rooms, as that is a separate item and will be treated separately under the title of "Ventilation".

Effect of Exposure on Heat Loss.—Too much emphasis cannot be laid on the effect of exposure on the heat loss from buildings. Experience has demonstrated that in a number of cases. There are numerous instances where a building in a protected position has been heated with steam or hot water and the results were most satisfactory, a temperature of 70° Fahrenheit being easily maintained inside of the building during zero weather. Those same buildings were then duplicated in other localities which were more exposed. The same owners, the same architects, the same builders and the same heating contractors in each case insured the buildings and the installations being identical. Yet, when subjected to the test of extremely windy or zero weather, the plants failed to heat properly.

It might be well to further point out that one heating plant cannot well be duplicated in another building having the same exposure, unless the duplications are carried to the most minute extreme, so far as radiators, boilers and pipes are concerned. For instance, if the original building had a boiler with a capacity of 1500 square feet of direct radiation of the five column 45 inch type, it would not do to use the same size boiler and the same size steam mains, then change from 1500 square feet of five column 45 inch radiation to the same number of square feet of 20 inch single column radiation. Owirg to the greater heat emissive capacity of the low type single column radiation, 1500 feet of it would be equal to over 2000 square feet of the larger size radiators, and the boilers and steam mains would not be of sufficient size to supply that amount.

It is generally supposed that windy weather cannot affect the heating of a building warmed with steam or hot water. That is a mistake, however, while a steam or hot water heated building is not affected, to so great an extent as one heated with hot air, still it is affected, and this is proven by the facts that the windward rooms are never as hot as the protected rooms, and that a building which is heated perfectly by a plant when located in a sheltered position, if exactly duplicated but exposed in a strong wind-swept location will not be heated properly with the same amount of radiation.

CHAPTER VIII

VENTILATION OF TURKISH BATHS

NECESSITY for Ventilation. — Thorough ventilation of a Turkish-bath establishment is of the utmost importance if the beneficial effect expected of a course of baths is to be realized. The dry heat so necessary for perfect evaporation from the person of the bathers cannot be maintained if a stream of dry air is not constantly passed through the sweat rooms to carry off the moisture exhaled and respired by sweat glands and lungs. Without ventilation the air of the hot rooms would soon become so charged with moisture that much of the benefit of a Turkish bath would be lost. In the other rooms of the establishment pure air is as desirable as in the sweat rooms. In the dressing room or separate resting room, where the bathers rest after a course of treatment in the baths, ventilation is absolutely necessary or the patrons will awake with a heavy, dull feeling due to lack of pure air, which will erroneously be blamed to the weakening or debilitating effects of the bath.

103

No less important than supplying dry air to carry off the moisture from the body and provide a plentiful supply of oxygen for the lungs is the removal of the organic matter, watery vapor and carbon dioxide thrown off by the lungs and skin of the bathers. Each person who undergoes the process of the bath rids his person of from several ounces to several pounds of organic matter in the hot rooms. This matter is as objectionable as other emanations from the body, and if allowed to remain in the rooms to accumulate and putrefy would make the compartments objectionable, to say the least.

Quantity of Fresh Air Required.—The quantity of fresh air required for ventilation can easily be foretold when the number of persons the bath is designed to accommodate simultaneously is known. The object of ventilation is to keep the air within the room as fresh and pure as the air outside. This is impossible in practice, however, or, at least, impracticable, owing to the size of ducts and apparatus required and the cost, so the reasonable and logical alternative is to keep the air within the rooms up to a certain standard of purity. To maintain this degree of purity in the sweat rooms of a Turkish bath at least 50 cubic feet of air per minute should be supplied for each occupant the rooms are designed to accommodate. The number of persons the sweat rooms will accommodate can be found, as previously explained, by allowing for each a floor space 4 feet by 6 feet. Another method of determining the amount of air required for ventilation is to allow for a certain number of changes per hour of the air within the rooms. This can easily be done by finding the cubical contents of the room and multiplying it by the number of air changes per hour. One change every six minutes, or ten changes per hour, will be about right for ordinary conditions.

Checking up those two methods for computing the amount of air required for ventilation it will be seen that they agree perfectly for the conditions given, and would be approximately the same for any other reasonable height of ceiling or size of floor space. Assuming a floor space of 4 feet by 6 feet and a height of ceiling of $12\frac{1}{2}$ feet would equal a cubical content of 300 cubic feet for each person. At one change every six minutes, or ten changes of air per hour, 3,000 cubic feet of air would be required for each person per hour, according to the air-change-per-cubical-contents method of computation. According to the air-per-person method 50 cubic feet of air per minute would be required, or $50 \times 60 = 3,000$ cubic feet per hour per person, and as each person would have allotted to him a floor space 4 feet by 6 feet it will be seen that the same amount of air will be required per person by both methods.

An allowance of 3,000 cubic feet of air per hour per person will generally be found sufficient for all rooms, even the hot rooms. No less supply than that, however, should be tolerated. In fact, if any change whatever is made in the quantity of air supplied it should be in favor of a larger, not a smaller amount, an allowance of 4,000 cubic feet per person per hour being a more desirable quantity.

Method of Ventilating Turkish Baths.— In order to move with regularity and certainty the large amount of air required for ventilation some mechanical means must be employed. Natural ventilation through flues and fireplaces, or ventilation by aspiration—that is, by means of coils or heaters placed in ventilator shafts—are not sufficiently reliable for the purpose. Further, the cost of ventilation by means of aspirating shafts would be far greater than the cost of moving the required amount of air by means of fans or blowers.

There are two methods of ventilating rooms by means of fans or blowers, which are known, respectively, as the exhaust method and the plenum method. In the exhaust method the fan, or blower, is installed in connection with the ventilating shafts, and the air is drawn from the rooms. This method is unsatisfactory for the reason that the operation of the fan creates a partial vacuum or lowers the pressure within the rooms which are to be ventilated and cold air leaks in around windows, doors and other openings, creating draughts within the apartments. Such a leakage inward is objectionable in ordinary buildings and doubly objectionable in the hot rooms of a Turkish bath, for the least draught of cold air in this highly heated atmosphere will be felt and seem excessively cold to the bathers. Further, with the exhaust method of ventilation there is no way to govern the quality or place of introduction of the air, which might be drawn from a contaminated source, thereby being as bad, or worse, than the air which is exhausted.

The plenum system is the best method of ventilation for the Turkish bath. In the plenum system air is drawn from the outside atmosphere at a selected source, heated by steam coils and forced, by means of a blower, into the several rooms to be ventilated, the excess air being forced out above the roof through specially constructed ventilator shafts. With this system the air within the rooms is under a slight pressure, consequently the leakage around windows, doors and other openings is outward, so that draughts will not be felt by the bathers; the temperature of the air and points of admission are completely under control, the action is positive, and the air can be heated to any desired temperature before passing into the rooms.

Example of Ventilating System.—The conventional arrangement of the ventilating apparatus and air-distributing system for a Turkish bath is shown in Fig. 25. Cold air is drawn through the screened inlet duct to the inside of a fresh-air chamber, where it passes successively through an air filter, a tempering coil, a fan or blower and a heating coil to the distributing ducts and flues. The arrangement shown in the illustration may be modified in many ways, and would be, according to the system of ventilation adopted. For instance, instead of forcing the air through the heating coils the fan or blower might draw the air through by suction and force it into the distributing ducts. In that case, of course, the tempering coil would be omitted. Again, if an air washer were used instead of a dry-air filter the washer would be placed between the tempering coils and the fan so the air

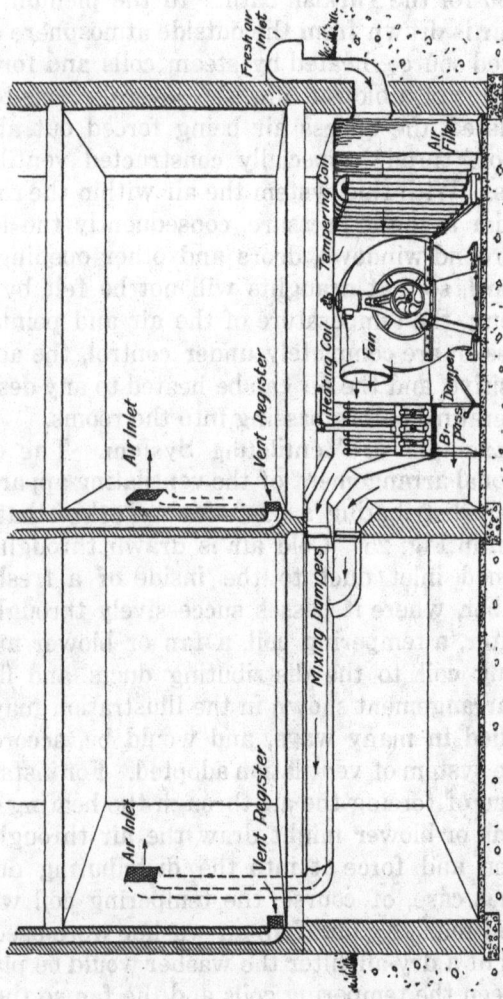

Fig. 25

would be slightly heated before passing through the washer, to prevent the water from being frozen in cold weather. Instead of a vertical steam engine a horizontal engine, either direct connected or belted, may be used for driving the fan, or instead of a steam engine an electric motor, either belted or direct connected, could be employed. Further, instead of a single duct system of air distribution the double duct system may, and no doubt would be used, while a disc or cone fan could be employed under some conditions instead of the centrifugul fan or blower indicated.

The illustration shows only the general arrangement and relative positions of the apparatus. The sizes, capacities and construction of the several parts will be described in detail later on.

Fresh-Air Inlet.—As the object of ventilation is to introduce pure air into the various compartments of the bathing establishment, the very first consideration should be to obtain the air in as pure a state as possible and free from dirt, dust, soot or other substances. This can best be done by locating the source of supply in an open space where there is a free circulation of air and placing the mouth of the intake several feet above the ground. In many cases the fresh-air supply is taken in through a screened window. Such a practice is not objectionable when the window is above the basement or ground floor, or even when taken in at the basement or ground-floor window if the yard outside is in every sense clean and the ground sodded or otherwise covered with a clean surface and free from dirt and traffic. Ordinarily, however, in business districts, where Turkish-bath establishments are generally located, the yard, court, alley, shaft or area from which the fresh-air inlet is to be taken is not wholly free from dust and dirt, not to mention contamination and vitiation of the air from other causes. In such cases the better practice is to extend the fresh-air inlet 18 or 20 feet above the ground level, or in some cases it might be desirable to locate it above the roof, in either case turning the inlet face downward, so that dust and dirt, snow or rain, will not settle therein, and cover the opening with a screen or grating to keep paper and other light substances from being drawn within by the current of air.

It might seem as though the point, or place, of intake would not matter so much when air filters or air washers are used. It should be borne in mind, however, that filters and washers remove only the mechanical impurities from the air; they do not purify it chemically by supplying oxygen which has been depleted or remove the excess of carbon dioxide or other deleterious gases. Even if it did there would be no logic in supplying air from a notoriously polluted source, depending upon mechanical apparatus to purify it, thereby placing upon the apparatus the extra burden which it would have in removing the mechanical impurities, such as dust, soot and dirt, which would be in excess over that in air taken at a more favorable point.

Sometimes, in order to place the fresh-air intake at a favorable point, it will necessitate the construction of brick or concrete ducts underground clear across the building from the point of intake to the fresh-air chamber. When such is the case manholes should be provided in the underground ducts to facilitate cleaning.

Size of Fresh-Air-Inlet Duct or Shaft.— The size of inlet duct for a building is determined by the total number of cubic feet of air required per minute for ventilation and the velocity at which the air will be delivered. In ordinary buildings—that is, those used for educational, business, medical and factory purposes—the velocity of air throughout the heating and ventilating system varies from 900 feet per minute, in some cases, to as high as 2,500 feet per minute in other cases. High velocities are sometimes necessary for economic reasons in order to keep down both the size and the cost of the distributing ducts, while in other cases high velocities are necessary to insure a thorough diffusion throughout the entire space to be heated and ventilated. In Turkish-bath design, however, high velocities must be avoided, and the highest rate at which air should pass through any of the ducts or flues is, perhaps, 900 feet per minute. If fresh-air-intake ducts, flues or shafts are proportioned to supply the total quantity of air required at this velocity they will not be found too small and will be large enough for any conditions which might arise. When the total number of cubic feet of air required to heat a Turkish bath has been calculated the area of duct in square inches required to conduct that amount of air at

111

TABLE VII

FLUE AREA REQUIRED FOR DIFFERENT VOLUMES AND VELOCITIES OF AIR

Volume of Air in Cubic Feet per Minute	Velocities in Feet per Minute														
	200	300	400	500	600	700	800	900	1,000	1,100	1,200	1,300	1,400	1,500	1,600
	Area in Square Inches of Flues Required														
100	72	48	36	29	24	21	18	16	14	13	12	11	10	9.6	9.
125	90	60	45	36	30	26	23	20	18	16	15	14	13	12.	11.3
150	108	72	54	43	36	31	27	24	22	20	18	16	15	14.4	13.5
175	126	84	63	50	42	36	32	28	25	23	21	19	18	16.8	15.8
200	144	96	72	58	48	41	36	32	29	26	24	22	21	19.2	18.
225	162	108	81	65	54	46	41	36	32	29	27	25	23	21.6	20.3
250	180	120	90	72	60	51	45	40	36	33	30	28	26	24	22.5
275	198	132	99	79	66	57	50	44	40	36	33	30	28	26.4	24.8
300	216	144	108	86	72	62	54	48	43	39	36	33	31	28.8	27.
325	234	156	117	94	78	67	59	52	47	43	39	36	33	31.2	29.3
350	252	168	126	101	84	72	63	56	50	46	42	39	36	33.6	31.5
375	270	180	135	108	90	77	68	60	54	49	45	42	39	36.	33.8
400	288	192	144	115	96	82	72	64	58	52	48	44	41	38.4	36.
425	306	204	153	122	102	87	77	68	61	56	51	47	44	40.8	38.3
450	324	216	162	130	108	93	81	72	65	59	54	50	46	43.2	40.5
475	342	228	171	137	114	98	86	76	68	62	57	53	49	45.6	42.8
500	360	240	180	144	120	103	90	80	72	65	60	55	51	48.	45.
525	378	252	189	151	126	108	95	84	76	69	63	58	54	50.4	47.3
550	396	264	198	158	132	113	99	88	79	72	66	61	57	52.8	49.5
575	414	276	207	166	138	118	104	92	83	75	69	64	59	55.2	51.8
600	432	288	216	173	144	123	108	96	86	79	72	66	62	57.6	54.
625	450	300	225	180	150	129	113	100	90	82	75	69	64	60.	56.3
650	468	312	234	187	156	134	117	104	94	85	78	72	67	62.4	58.5
675	486	324	243	194	162	139	122	108	97	88	81	75	69	64.8	60.8
700	504	336	252	202	168	144	126	112	101	92	84	78	72	67.2	63.
725	522	348	261	209	174	149	131	116	104	95	87	80	75	69.6	65.3
750	540	360	270	216	180	154	135	120	108	98	90	83	77	72.	67.5
775	558	372	279	223	186	159	140	124	112	101	93	86	80	74.4	69.8
800	576	384	288	230	192	165	144	128	115	105	96	89	82	76.8	72.
825	594	396	297	238	198	170	149	132	119	108	99	91	85	79.2	74.3
850	612	408	306	245	204	175	153	136	122	111	102	94	87	81.6	76.5
875	630	420	315	252	210	180	158	140	126	115	105	97	90	84.	78.8
900	648	432	324	259	216	185	162	144	130	118	108	100	93	86.4	81.
925	666	444	333	266	222	190	167	148	133	121	111	102	95	88.8	83.3
950	684	456	342	274	228	195	171	152	137	124	114	105	98	91.2	85.5
975	702	468	351	281	234	201	176	156	140	128	117	108	100	93.6	87.8
1,000	720	480	360	288	240	206	180	160	144	131	120	111	103	96.	90.

112

any velocity from 200 feet per minute to 1,600 feet per minute can be found in Table VII.

Table VII does not give areas for volumes below 100 cubic feet per minute or above 1,000 cubic feet per minute. If areas for quantities larger or smaller than these amounts are required they can be determined readily by multiplying the volume of air to be moved by the area required for the given velocity. For instance, if 6,525 cubic feet of air are to be delivered at a velocity of 900 feet per minute, by multiplying the area (160 square inches) required to move 1,000 cubic feet at a rate of 900 feet per minute by 6.5, which is the number of 1,000 volumes to be moved, will give 160 × 6.5, or 1,040 square inches area required in the pipe.

If on the other hand, the volume of air to be moved be less than 100 feet, for instance, .65 cubic feet per minute at a velocity of 900 feet per minute, the area (16 square inches) required to deliver 100 cubic feet at that velocity would be multiplied by the quantity of air to be delivered. Thus, 16 × .65 = 10.4 square inches area in the pipe. In this case, of course, the quantity .65 would be treated as a decimal, as it is only 65-100 of the unit amount multiplied. In like manner, when quantities containing odd parts of thousands are multiplied the odd parts must be pointed off as decimal places.

Having the area of pipe required in square inches the actual dimensions of the pipe, either for round pipe or square, can be determined by means of Table VIII.

TABLE VIII.—SIZE OF ROUND AND SQUARE PIPE FOR GIVEN AREAS

Area of Pipe in Square Inches	Diameter of Round Pipe in Inches Having Same Area	Sides of Square Pipe of Same Area in Inches	Area of Pipe in Square Inches	Diameter of Round Pipe in Inches Having Same Area	Sides of Square Pipe of Same Area in Inches	Area of Pipe in Square Inches	Diameter of Round Pipe in Inches Having Same Area	Sides of Square Pipe of Same Area in Inches
.7854	1.	.89	346 36	21.	18.61	1,320.26	41.	36 34
1.767	1.5	1.33	363 05	21.5	19.05	1,352.66	41.5	36.78
3.1412	2.	1.77	380 13	22.	19 50	1,385 45	42.	37.22
4.909	2.5	2.22	397.61	22.5	19 94	1,418.63	42.5	37 66
7.069	3.	2.66	415 48	23.	20 38	1,452 20	43.	38 11
9.612	3.5	3.10	433 74	23.5	20 83	1,486 17	43.5	38 55
12.566	4.	3.54	452 39	24.	21 27	1,520 53	44.	38 99
15.904	4.5	3.99	471 44	24.5	21 71	1,555.29	44.5	39 44
19.635	5.	4.43	490 88	25.	22 16	1,590 43	45.	39 88
23.758	5.5	4.87	510 71	25.5	22 60	1,625 97	45.5	40 32
28 274	6.	5.32	530 93	26.	23 04	1,666 91	46.	40.77
33.183	6.5	5.76	551 55	26.5	23 49	1,698 23	46.5	41 12
38.185	7.	6.20	572 56	27.	23 93	1,734 96	47.	41.65
44.179	7.5	6.65	593 96	27.5	24 37	1,772 06	47.5	42.10
50.266	8.	7.09	615 75	28.	24 81	1,809 56	48.	42.56
56.745	8.5	7.53	637 94	28.5	25 26	1,847 46	48.5	42.98
63.617	9.	7.98	660 52	29.	25 70	1,885 75	49.	43.43
70.882	9.5	8.42	683 49	29.5	26 14	1,924 43	49.5	43.87
78.540	10.	8.86	706 86	30.	26 59	1,963 50	50.	44.31
86.590	10.5	9.30	730 62	30.5	27 03	2,002 97	50.5	44.75
95.03	11.	9 75	754 77	31.	27 47	2,042 83	51.	45 20
103.87	11.5	10 19	779 31	31.5	27 92	2,083 08	51 5	45.64
113.10	12.	10 63	804 25	32.	28 36	2,123 72	52.	46.08
122.72	12.5	11 08	829 58	32.5	28 80	2,164 76	52.5	46.53
132.73	13.	11 52	855 30	33.	29 25	2,206 19	53.	46.97
143 14	13.5	11 96	881 41	33.5	29 69	2,248 01	53.5	47.41
153.94	14.	12.41	907 92	34.	30 13	2,290 23	54.	47.86
165.13	14.5	12.85	934 82	34.5	30 57	2,332 83	54.5	48 30
176.72	15.	13 29	962 11	35.	31 02	2,375.83	55.	48.74
188.69	15.5	13 74	989 80	35.5	31 46	2,419.23	55.5	49 19
201 06	16.	14 18	1,017 88	36.	31 90	2,463.01	56.	49 63
213.83	16.5	14 62	1,046 35	36.5	32 35	2,507 19	56.5	50 07
226.98	17.	15 07	1,075 21	37.	32 79	2,551 76	57.	50 51
240.53	17.5	15 51	1,104 47	37.5	33 23	2,596 73	57 5	50 96
254.47	18.	15.95	1,134 12	38.	33 68	2,642 09	58.	51 40
268.80	18.5	16.40	1,164 16	38.5	34 12	2,687 84	58.5	51 54
283.53	19.	16.84	1,194 59	39.	34 56	2,733 98	59.	52 29
298.65	19 5	17.28	1,225.42	39.5	35 01	2,780 51	59.5	52 73
314 16	20.	17.72	1,256 64	40.	35 45	2,827 44	60.	53 17
350.06	20 5	18.17	1,288 25	40.5	35.89	2,874 76	60 5	53 62

Suppose, for instance, that a pipe containing an area of 1,040 square inches is required. Glancing down the column of areas to 1,046, which is the nearest value, it will be found that a round pipe having an inside diameter of 36½ inches would contain that amount of area, and a square pipe having sides 32⅓ inches an equal area.

Air Filters – From whatever source the supply of air for ventilation is obtained there will be more or less soot, dust, smoke and other impurities contained therein. If the velocity at intake is very high many particles of light floating matter will be drawn toward the mouth of the duct by the current of air rushing in, and, unless removed, will be discharged into the rooms where the bathers are assembled, to soil their persons after being cleansed. It might seem as though the small amount of dust or dirt in a unit quantity of air would be too insignificant for consideration, but it must be borne in mind that the unit quantity is multiplied time and time again, so that the insignificant amount of suspended matter becomes an important and noticeable quantity, sufficient to make dusty the rooms of a bathing establishment if the air is not filtered or washed before being discharged into the bathing compartments.

A "Whitley" apparatus, designed for filtering air before it is delivered to the fan, is shown in Fig. 26. In the illustration the filter is shown housed in a sheet-metal casing, although it may be enclosed by masonry walls, wood, or any other suitable material.

The drawing shows the folds or convolutions of the filtering material in a vertical position. Instead of vertical folds, however, the folds of the fabric may be horizontal, as would be necessary

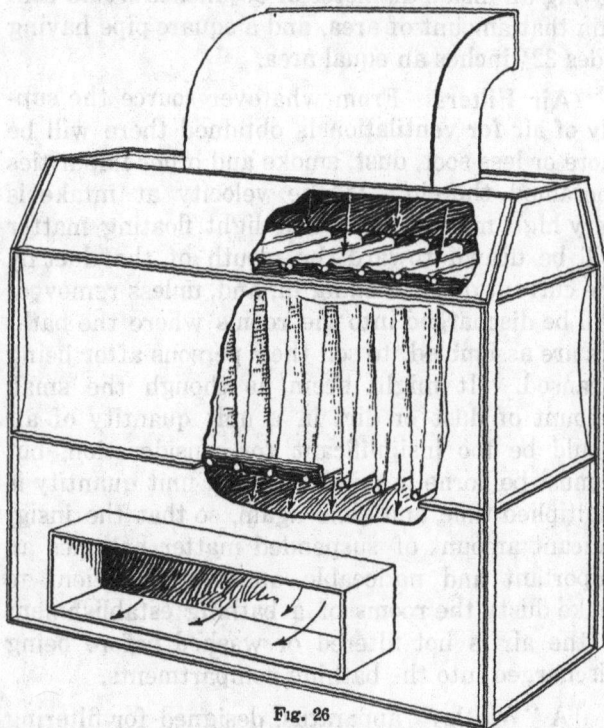

Fig. 26

Air Filter

if a filter of this type were housed in a horizontal flue.

The principle of the filter consists in interposing a coarse woven fabric between the inlet to the fresh-air chamber and the fan, so that the

116

mechanical impurities will be strained out, as it were, before the air enters the distributing systems. As the fabric will interpose considerable resistence to the flow of air through its meshes the surface of the filtering material must be much larger than the area of the inlet duct to allow for the flow of a sufficient volume of air without excessive friction.

Ordinarily, in this type of filter an allowance of 100 to 1 is made. That is, the area of filter surface is one hundred times greater than the area of the fresh-air-inlet shaft or duct. In order to get this proportionate area of filter surface in a small space the fabric is folded back and forth as indicated in the drawing.

In operation the filtering material must be cleaned occasionally or the meshes of the fabric will become so clogged with dirt that air will not pass through in sufficient volume for efficient ventilation.

The fabric used for an air filter must be of coarse mesh, like cheesecloth or burlap, but of tough fiber, to stand the strain of air passing through and the cutting and tearing effect of the particles it stops.

Air Washers.—Wet filters, or air washers, are more extensively used in mechanical ventilation than are dry filters of types similar to the one just described, for the reason that air washers not only purify the air of all dirt and dust, but impart much needed moisture or humidity to it at the same time. This is important in all classes of buildings except Turkish-bath establishments, for a humid air is required for efficient ventilation. In the hot rooms of a Turkish bath, on the contrary, a dry atmosphere with high evaporative capacity is desired, and the addition of moisture to the air is not beneficial. For all other rooms throughout the building, however, an atmosphere with the right degree of humidity should be maintained. These different requirements of the various compartments of a Turkish bath should be kept in mind when designing the ventilating system, so that air for each room will be supplied not only in sufficient quantity but likewise of suitable quality. If an air washer is used for removing the dirt and dust from the air used for ventilation some means should be employed to remove all the moisture possible from the air before it reaches the hot rooms. Four moisture extractors instead of the usual three would probably reduce the excess moisture to such a degree that while it would be suitable for all other compartments in the establishment the relative humidity in the hot rooms would be sufficiently low when the temperature of the air had been raised to the required degree. On the other hand, if a dry filter be used means must be provided to furnish humidity to those compartments throughout the building where a normal or relatively humid atmosphere is desired.

A Peerless air washer is shown in Fig. 27. The entire apparatus consists of a tempering coil, spray chamber with spray nozzles, moisture extractors and centrifugal pump. The pump is belt-connected to the engine which drives the fan, so that when the fan is in operation the pump and

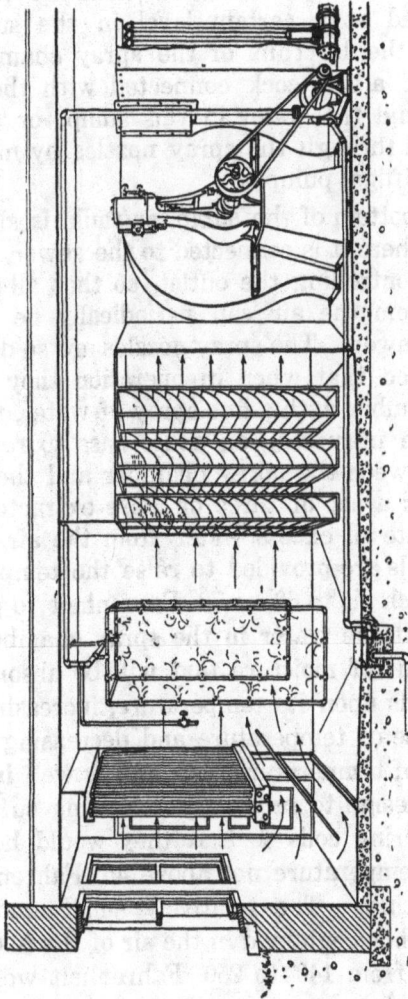

Fig. 27
Peerless Air Washer

air washer will be in operation also. Water is maintained at a certain level in the sump, or tank, at the bottom of the spray chamber, by means of a ball cock connected with the water supply, and the water in this sump, or tank, is circulated through the spray nozzles by means of the centrifugal pump.

The bottom of the sump, or tank, is sloped to a point where it is connected to the sewer, a valve or gate controlling the outlet, so that all dirt removed from the air can periodically be washed into the sewer. The spray nozzles are so designed and located that when in operation they fill the spray chamber with a fine spray of water, through which the in-rushing air must pass to reach the fan. Between the spray chamber and the fan is interposed a set of three or more extractor units to eliminate all excess water from the air. Tempering coils are provided to raise the temperature of the fresh air to 60° or 70° Fahrenheit to prevent freezing of the water in the spray chamber. As the amount of moisture that will be absorbed by air depends upon its temperature, increasing with an increase of temperature and decreasing with a decrease of temperature, it might be well in Turkish-bath design to provide less heating surface in the tempering coils so that they would heat the air to a temperature not above 45° Fahrenheit in zero weather. The relatively small amount of moisture then contained in the air of the hot rooms heated to from 140° to 250° Fahrenheit would not be noticeable.

CHAPTER IX

FANS OR BLOWERS

SELECTION of a Fan.—There are only two types of fans which are suitable for ventilation purposes in Turkish bath buildings. One is the centrifugal fan, the other is the cone fan. A centrifugal fan with bottom discharge and full housing of metal is shown in Fig. 28. This is the design commonly adopted for the small and medium fans, such as would be required in a moderate-sized establishment. The fan in this illustration is direct connected to an upright vertical engine, making what is called a steam fan. Steam fans may be had with any conceivable style of housing, with the discharge outlet at any convenient point in the casing, and driven by any kind of prime mover. When designing the ventilation system the best way is to consult the catalogues of fan manufacturers to find the design of fan suitable for the purpose.

A cone fan is shown in Fig. 29. This fan is illustrated with a pulley for belt drive, although, like the centrifugal fan, it may be driven by a belt, steam engine or electric motor, care being taken to keep the apparatus out of the way of air supply and delivery.

Cone fans and centrifugal fans each have their own field of usefulness. In small establishments, where the air delivery ducts are direct, of large area and short runs, so there will be but little frictional resistance offered to the flow of air, the cone fan will prove the more satisfactory, as its operation is much less noisy than that of the centrifugal fan. On the other hand, in large establishments, or in any kind of an establishment where the air must be forced a considerable distance through ducts and flues, the centrifugal fan will prove the more efficient, owing to its better delivery against pressure.

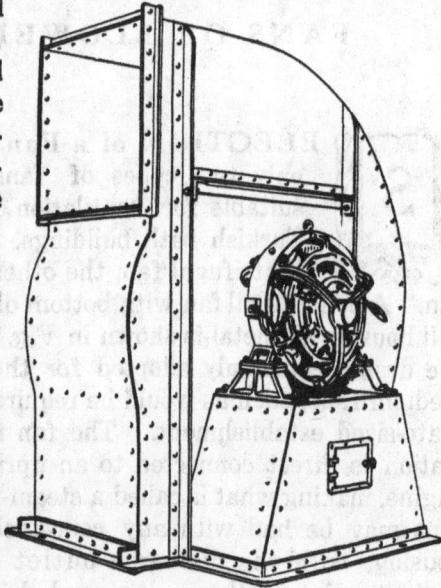

Fig. 28

Centrifugal Fan

122

When centrifugal fans must be located close to the public rooms of a Turkish bath a canvas accordion-fold section can be installed between the fan and the main duct, to overcome the effect of vibrations which otherwise cannot be silenced. Under no consideration should the engine be located inside of the fresh-air room or in close proximity to the air ducts or the throb of the engine can be heard all through the building, a condition which would be very annoying to many of the patrons.

When the centrifugal fan is to be used the steam-heating coils may be so located with reference to the rest of the apparatus that the air either will be drawn through the coils to the fan or so it will be forced through after leaving the fan. With the cone fan, however, the air must be drawn through the coils; it cannot successfully be forced

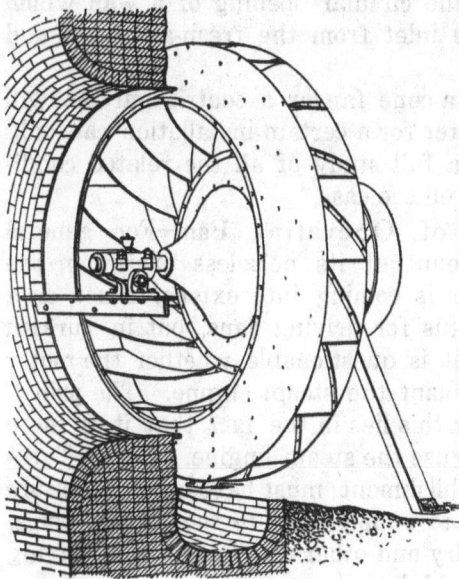

Fig. 29
Cone Fan

123

through. Cone fans are suitable only for delivering large volumes of air at low velocities, a condition which obtains in Turkish bath ventilation, where a speed of 1,000 feet per minute is about the highest that will be required in any part of the system. They can be set without casings and are best set in the circular opening of a wall which serves as the inlet from the fresh-air shaft and coil room.

Whether a cone fan or a centrifugal fan will prove the better for a certain installation can only be told after a full study of all the related conditions bearing on the case.

Method of Operating Fan—For general work, on account of its noiseless operation, the electric motor is coming into extensive use as a motor apparatus for driving fans, but in Turkish bath design it is questionable whether the motor will ever supplant the steam engine. The principal reason for this lies in the fact that it is more economical to use the steam engine. Every Turkish bath establishment must have a high-pressure steam plant to supply heat to the sweat rooms, operate laundry and other machinery, and, having the available high pressure, true economy lies in using it, as the exhaust steam from the engine can be used for heating the tempering coils.

The exhaust steam may likewise be used for heating the main-heater coils, or as many sections of them as there is exhaust steam to supply, if the air in the ventilation system is simply raised to the temperature required for the dressing and other rooms of the building having normal temperatures, and the air for the sweat rooms raised to their respective temperatures by means of reheaters. This is, no doubt, the method which would be adopted, so that steam engines would naturally be the best prime movers for ventilation fans.

The type of steam engine, whether upright or horizontal, direct connected or pulleyed for belt drive, must be determined in each case from all data available. Ordinarily, where quietness of operation is not essential—as, for instance, when the engine and fan are remote from the bath rooms—a direct connected upright engine will prove the more satisfactory, as it will cost less, occupy less space and be less likely to get out of order than a belt-connected engine. On the other hand, when the fan and engine are so situated or located that noiselessness of operation is essential, a horizontal-engine belt connected to the fan should be provided.

When a horizontal engine with belt pulley is installed an auxiliary pulley for operating the centrifugal pump that circulates water through the air washer can be provided.

Speed of Fans.—The speed at which fans can be operated in Turkish bath ventilators depends upon the greatest speed at which a fan can be operated without making an objectionable noise. This limit has been found to be a peripheral velocity of about 3,500 feet per minute for straight-blade fans, and about 4,000 feet per minute for multivane fans, and in order to be well within the noise-producing velocity it is well not to operate

125

the fans at a greater speed than from 2,800 feet per minute for straight-blade fans, and 3,500 feet per minute for multivane fans. Indeed, greater economy will result from using larger fans, run at slow speed, than small fans run at high speed, to deliver a given volume of air, while at the same time a smoother-running, less-noisy apparatus will result. It must be borne in mind, however, that true economy does not lie in going to the other extreme and installing a huge fan to run at an extremely low speed. No matter what size fan is used it must be run at sufficient speed to overcome the resistance due to friction in the ducts and flues and at the same time maintain a sufficient pressure to move the air along to the rooms. This excess pressure must be very slight, however, or too high velocity will be established throughout the distributing system. The fans should never be calculated to run at a velocity which would produce a greater pressure than that required to balance a column of water one-half inch high, which is equal to a pressure of 0.29 ounce per square inch.

The velocities produced in a duct 100 feet long and 1 foot in diameter by various pressures in excess of the atmosphere can be found in Table IX.

This table was calculated for a barometric pressure of 30 inches and a temperature of 60° Fahrenheit. For other temperature the results would have to be multiplied by suitable factors, but as 60° Fahrenheit is about the temperature at which the air would ordinarily reach the fan for ventilation purposes the values in the table may

TABLE IX.*
VOLUME OF AIR DISCHARGED AT VARIOUS PRESSURES

Difference of Pressure		Velocity in Feet per Second Through Round Pipe 100 Feet Long and One Foot in Diameter	
Inches of Water	Ounces per Square Inch	By Accurate Formula	By Appromimate Formula, $l_1 = \sqrt{2} - gh$, Using a Coefficient of 0.7
.01	.006	4.3	4.6
.05	.030	9.6	9.5
.1	.058	14.5	14.5
.2	.116	19.4	20.5
.3	.174	23.6	25.1
.4	.232	27.4	29.1
.5	.289	30.5	32.5
.6	.347	34.0	35.2
.7	.405	36.0	38.3
.8	.463	39.2	40.7
.9	.512	41 0	43.7
1.0	.579	43.0	45.7
2.0	1.158	61.1	65.2
3.0	1.303	78.0	78.2
4.0	2.316	85.3	91.1
5.0	2.895	86.2	103.3
6.0	3.474	104.0	113.3
7.0	4.053	114.0	122.1
8.0	4.622	121.0	130.6
9.0	5.221	128.0	138.8
10.0	5.790	136.0	145.7
11.0	6.369	142.0	153.0
12.0	6.948	143.0	159.6

*Carpenter.

be taken as given and will be found sufficiently correct.

Size and Capacity of Centrifugal Fans.— The size and capacity of fan required for ventilation purposes will depend always upon the volume of air to be delivered per minute. When that quantity is determined the selection of a fan which will deliver that amount of air is more a matter of judicious selection from manufacturers' catalogues than a matter of calculation. The theory of design of fans is of no importance to the architect, for there are various stock types and sizes which can be had and the variety of size and design is so large that a stock apparatus can be had for almost any conceivable use. All that is necessary is to select one having a sufficiently high rating to deliver the amount of air required at a low velocity. In selecting a fan from manufacturers' catalogues, however, the designer must not lose sight of the fact that manufacturers' ratings are generally based on the free discharge of air into the atmosphere. In actual practice, forcing air through a system of ducts and flues, a fan will deliver only from 50 to 65 per cent. of this theoretical quantity, depending on the length and size of the ducts. Ordinarly, in selecting a fan for ventilation purposes, it is assumed that the fan will deliver only 50 per cent. of its rated capacity, and one is specified which has a capacity just double that required.

The capacities at different revolutions and air pressures, also the dimensions and horse-power of Sturtevant centrifugal straight-blade fans, can be found in Table X.

A careful examination of the table will show the economy of using large fans operated at moderate speed over smaller fans operated at greater speed. Take, for example, the discharge of 50,000 cubic feet of air per hour into the atmosphere. By glancing at the table it will be seen that a 5½ by 3 foot fan, operated at a velocity of 300 revolu-

TABLE X.—CAPACITIES OF STRAIGHT-BLADE CENTRIFUGAL FANS

Size in Feet	Diameter of Fan in Inches	Diameter of Inlet in Inches	Width of Housing in Inches	Pulley		¼-Ounce Pressure or 0.43 Inch Water			½-Ounce Pressure or 0 68 Inch Water			¾-Ounce Pressure or 1 3 Inches Water			1-Ounce Pressure or 1 73 Inches Water		
				Diameter	Face	Revolutions per Minute	Cubic Feet per Minute	Horse Power	Revolutions per Minute	Cubic Feet per Minute	Horse Power	Revolutions per Minute	Cubic Feet per Minute	Horse Power	Revolutions per Minute	Cubic Feet per Minute	Horse Power
5½x3	66	47	36	22	9½	150	24,734	1.2	212	34,992	3.4	259	32.892	5 6	300	49,524	9 7
6 x3½	72	52¼	42	24	10½	138	31,694	1.5	194	44,857	4.4	238	54 966	7.1	275	63,463	12.4
7 x4	84	60½	48	28	12½	118	42,167	2.1	166	59,681	5 8	204	73.130	9 5	235	84,436	16 5
8 x4	96	68½	48	32	12½	103	47,486	2.3	145	67,218	6.6	178	82 355	10.8	206	95,086	18 6
9 x4½	108	77	54	36	12½	91	60,992	3.0	129	86,180	8 4	159	105.602	13.7	183	121,938	23.8
10 x5	120	85½	60	42	12½	82	75,816	3.7	116	107,313	10.5	143	131.475	17.1	165	151,800	29.6
12 x6	144	102½	72	48	14½	69	108,703	5.3	97	153,850	15 0	119	188.520	24.5	137	217,340	42.4
14 x7	168	120	84	54	14½	59	149,840	7 3	83	212,070	20.7	102	259 765	33.8	118	300,000	58.4
15 x7½	180	128¼	90	60	16½	55	172,740	8 4	78	244,487	23.8	95	299.584	39.0	110	345,900	67.4

TABLE XI.—SIZES AND CAPACITIES OF STURTEVANT CENTRIFUGAL MULTIVANE FANS

No. or Size of Fan	Resistance	Exhaust or Ordinary Ventilating Systems Without Heater or Air Screens — R.P.M.	— Vol. at High Mech. Eff.	— Brake H.P.	Drying with Heaters — R.P.M.	— Vol. at High Mech. Eff.	— Brake H.P.	Public Building with Heaters — R.P.M.	— Vol. at High Mech. Eff.	— Brake H.P.	Factory with Heaters — R.P.M.	— Vol. at High Mech. Eff.	— Brake H.P.	Roundhouse with Heaters — R.P.M.	— Vol. at High Mech. Eff.	— Brake H.P.	Induced Draft 18 Lbs. Coal — R.P.M.	— Vol. of Gases 500°F.	— Brake H.P.	25 Lbs. Coal — R.P.M.	— Vol. of Gases 500°F.	— Brake H.P.	40 Lbs. Coal — R.P.M.	— Vol. of Gases 500°F.	— Brake H.P.
3	Min.	290	587	.03	410	825	.08	580	1,170	.23	649	1,310	.32	768	1,550	.53	1,100	2,220	.86	1,290	2,600	1.40	1,454	2,940	2.00
3	Max.	348	905	.06	492	1,275	.16	696	1,805	.45	779	2,020	.63	921	2,390	1.04	1,320	3,430	1.70	1,548	4,020	2.75	1,745	4,530	3.94
4	Min.	242	845	.04	341	1,190	.12	483	1,690	.33	540	1,890	.46	639	2,240	.76	917	3,210	1.25	1,072	3,750	2.01	1,210	4,250	2.90
4	Max.	290	1,305	.08	409	1,840	.23	580	2,610	.65	648	2,920	.91	766	3,450	1.50	1,100	4,950	2.45	1,286	5,790	3.96	1,451	6,550	5.70
5	Min.	207	1,150	.06	292	1,630	.16	414	2,300	.45	455	2,530	.62	548	3,050	1.04	784	4,360	2.22	920	5,110	3.57	1,036	5,770	3.93
5	Max.	248	1,770	.11	351	2,510	.31	497	3,550	.88	545	3,900	1.21	657	4,700	2.04	940	6,720	3.34	1,102	7,890	5.39	1,243	8,900	7.73
6	Min.	181	1,500	.07	256	2,120	.21	363	3,000	.58	480	3,350	.81	513	3,940	1.35	688	5,670	2.65	806	6,650	4.07	909	7,510	5.14
6	Max.	217	2,310	.14	307	3,260	.40	435	4,620	1.15	575	5,160	1.60	615	6,110	2.66	825	8,750	5.21	966	10,260	8.00	1,090	11,590	10.10
7	Min.	145	2,340	.11	205	3,310	.32	290	4,680	.91	324	5,230	1.28	384	6,200	2.12	549	9,000	3.51	645	10,560	5.55	727	11,900	8.09
7	Max.	174	3,610	.22	246	5,100	.63	348	7,220	1.79	389	8,070	2.51	460	9,550	4.16	659	13,870	6.90	773	16,250	10.90	872	18,350	15.90
8	Min.	121	3,370	.16	171	4,770	.32	242	6,750	1.81	270	7,540	1.84	320	8,940	3.04	458	12,800	4.97	537	15,000	8.04	605	16,900	11.50
8	Max.	145	5,200	.32	205	7,350	.63	290	10,400	2.58	324	11,620	2.51	342	13,770	5.95	550	19,700	9.78	645	23,100	15.80	726	26,100	12.60
9	Min.	104	4,610	.22	147	6,480	.63	208	9,200	1.80	232	10,250	2.50	274	12,100	4.13	496	17,400	6.76	561	20,440	10.90	648	23,000	15.80
9	Max.	125	7,110	.44	176	10,000	1.24	232	14,200	3.53	278	15,800	4.91	329	18,000	8.13	552	26,800	13.30	673	31,400	21.40	668	35,500	31.00

Group	Stat	1	2	3	4	5	6	7	8	9	10	11	12	13	14	15	16	17	18	19	20	21	22	23	24	25
10	Min.	91	6,020	.30	128	8,500	.83	204	182	12,000	2.35	204	13,500	3.29	240	15,900	5.39	344	22,800	8.85	403	26,700	14.30	457	30,300	20.70
10	Max.	-09	9,280	.58	154	13,100	1.63	244	218	18,550	4.61	240	20,800	6.47	288	24,500	10.60	412	35,100	17.40	483	41,200	28.10	548	46,700	40.70
11	Min.	81	7,670	.37	114	10,880	1.05	180	161	15,200	2.97	180	17,100	4.16	216	20,200	6.86	305	28,900	11.30	357	33,900	18.20	403	38,300	26.10
11	Max.	97	11,830	.73	137	16,700	2.07	216	193	23,500	5.84	216	26,350	8.19	255	31,100	13.50	366	44,600	22.10	428	52,200	35.80	454	59,000	51.30
12	Min.	73	9,400	.46	103	13,270	1.29	163	145	18,700	3.65	163	21,000	5.08	192	24,700	8.44	275	35,500	13.80	323	41,700	22.40	364	47,100	32.20
12	Max.	87	14,500	.90	123	20,450	2.54	195	174	28,900	7.18	195	32,400	10.00	230	38,200	16.60	330	54,800	27.20	387	64,300	44.00	437	72,600	63.20
13	Min.	66	11,850	.55	94	16,100	1.57	148	132	22,700	4.42	148	25,400	6.15	174	30,000	10.30	250	43,100	16.80	298	50,500	27.10	331	57,000	38.90
13	Max.	79	17,500	1.09	112	24,800	3.08	177	158	35,000	8.69	177	39,200	12.10	209	46,300	20.20	300	66,500	33.00	351	77,900	53.20	397	87,900	76.50
14	Min.	61	13,600	.66	86	19,200	1.87	136	121	27,000	5.23	136	30,400	7.37	160	35,800	12.20	229	51,200	19.90	269	60,300	32.30	304	68,000	46.30
14	Max.	73	21,000	1.30	103	29,600	3.68	163	145	41,700	10.30	163	47,000	14.50	192	55,200	24.00	275	79,000	39.20	323	93,000	63.50	364	104,800	91.10
15	Min.	56	15,800	.77	79	22,400	2.19	125	112	31,600	6.15	125	35,400	8.65	148	41,800	14.30	213	60,300	23.50	249	70,400	37.80	280	79,500	54.40
15	Max.	67	24,400	1.51	95	34,600	4.30	148	134	48,800	12.10	148	54,700	17.00	177	64,500	28.10	255	93,000	46.10	298	108,500	74.20	336	122,600	107.00
16	Min.	52	18,400	.89	74	26,200	2.54	115	103	36,800	7.17	115	41,300	10.00	139	48,600	16.60	197	70,000	27.30	230	82,000	44.00	259	92,200	63.00
16	Max.	62	28,400	1.76	88	40,300	5.00	139	124	56,700	14.10	139	63,600	19.70	164	75,000	32.60	236	108,000	53.60	276	126,400	86.50	311	142,200	124.00
17	Min.	49	21,100	1.03	69	29,800	2.91	108	97	42,200	8.24	108	47,300	11.60	130	56,100	19.10	184	79,800	31.10	216	94,000	50.60	243	105,700	72.20
17	Max.	58	32,500	2.02	82	46,000	5.72	130	116	73,000	16.20	130	86,500	22.70	156	86,500	37.60	235	131,500	61.20	259	145,000	99.40	291	163,000	142.00

If fans of double inlet type are used the Volume is increased 12 per cent. and Brake Horse Power 20 per cent. for any case in tables.

Volumes above (except Induced Draft) are for fans blowing through heaters. If fans draw through heaters add 5 per cent. to R. P. M. and B. H. P.

tions per minute, will discharge 49,524 cubic feet; a 6 by 3½ foot fan will discharge 54,966 cubic feet, when operating at a speed of 238 revolutions per minute; a 7 by 4 foot fan will discharge almost 60,000 cubic feet per minute when operating at a speed of 166 revolutions per minute, and an 8 by 4 foot fan will discharge 47,486 cubic feet, at a velocity of 103 revolutions per minute. In the case of the 5½-foot fan, however, it requires 9.7 horse power to operate it at the speed of 300 revolutions per minute, while the 6-foot fan will deliver approximately the same quantity of air with an expenditure of only 7.1 horse power; the 7 foot with a consumption of 5.8 horse power; the 8 foot with 2.3 horse power. So far as horse power is concerned, therefore, economy would lie in using a large-size fan run at a comparatively low rate of speed. There are other reasons, however, which necessitate the use of large fans at slow speed in Turkish-bath ventilation. As was previously pointed out, in order to prevent noisy operation of the apparatus the fan must be run at a less speed than 3,500 feet per minute, and preferably not at greater speed than 2,800 feet per minute. This requirement would eliminate the two smaller fans mentioned, which run at peripheral velocities respectively of 5,484 and 4,485 feet per minute to deliver the required volume of air, while the large fans run at speeds of 3,633 and 2,588 feet respectively. There is a further consideration, also, which eliminates those two fans. As formerly stated, a fan should be operated so as not to produce a greater pressure than ½ ounce, with preferably ¼-ounce pressure, and both of the smaller fans above mentioned, which were run at speeds of 238 and 300 feet per minute, produced pressures respectively of ¾ ounce and 1 ounce per square inch.

Multivane Fans.—Multivane fans of the centrifugal type are becoming extensively used for ventilation purposes, as they are more efficient and less noisy than straight-blade fans when delivering equal volumes of air. The sizes, capacities, uses and various data regarding the Sturtevant multivane fans may be seen in Table XI. The size of fan listed in the first column is simply a size number and does not refer to the diameter of fan.

Size and Capacities of Cone Fans.—The capacities of cone fans run at various speeds are given in Table XII. In this table the quantities of air delivered are for actual working conditions through ventilation ducts and flues and at pressures ranging from about 3-16 ounce to less than ¾ ounce per square inch. By maintaining a speed midway between the two extremes stated would give a suitable velocity for the purpose required.

TABLE XII.—SIZE AND CAPACITY OF CONE FANS

Diameter of Fan in Feet	Revolutions Per Minute	Delivery in Cubic Feet Per Minute	Pressure in Inches of Water	Horse Power
4	250 to 500	6,400 to 12,800	.31 to 1.23	3
5	200 to 400	10,000 to 20,000	.31 to 1.23	5
6	167 to 333	14,400 to 28,800	.31 to 1 23	7½
7	140 to 280	20,000 to 40,000	.31 to 1.23	10
8	125 to 250	25,500 to 51,000	.31 to 1.23	15
10	100 to 200	40,000 to 80,000	.31 to 1.23	20
12	83 to 167	57,500 to 115,000	.31 to 1.23	25

Cone fans must always be so installed that they can draw air through the heater coils and discharge it into a plenum chamber, from which the various ducts are taken. Owing to the low pressure against which this type of fan is designed to operate it cannot very well blow the air for ventilation through the heater coils. Further, it is desirable, in case an electric motor is used, to have the fan and motor on the house side of the heater, so the parts will he warm and the oil for the bearings in a free condition to run. Cone fans with direct-connected motors are quite noiseless in operation. When direct-connected motors are used they should be both dust proof and moisture proof. Electric motors for fans are generally of the direct-current type, wound for 110, 220 or 500 volts. They may be had direct connected to the fans, connected by means of belts or by means of quiet-running chains. When a motor is to be connected to a fan by belting or chain the motor should be mounted on a sliding bed, so that slack in the belt can be taken up by slightly shifting the motor.

COILS AND HEATERS

Main-Heater Coils.—The most satisfactory and efficient form of main heater for heating and ventilating purposes is a pipe-coil heater made up of 1-inch wrought pipe with cast-iron steam fittings, and screwed into cast-iron base sections, as shown in Fig. 30. In making up a pipe-coil heater the vertical runs of pipe are staggered so that air will have to take a zigzag course through the coil, thus bringing all the air into direct contact with the heating surface. The coil bases rest on an angle-iron foundation, which is provided at the end opposite the steam inlet with a series of expansion balls or rollers, which permit of easy movement of the base sections under the variations due to expansion and contraction.

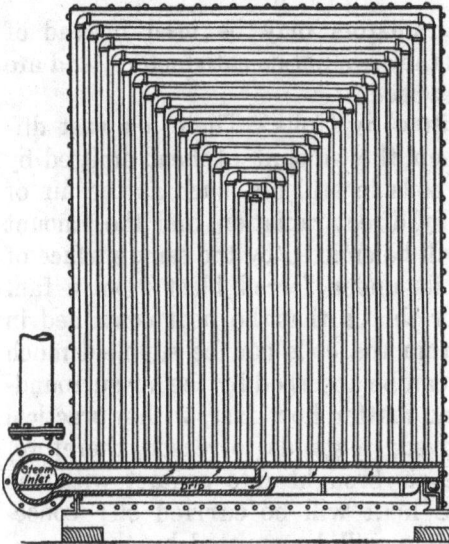

Fig. 30
Pipe Coil Heater

The steam coils should be divided into sections, which can be shut off or turned on, according to the requirements of the weather, and when steam engines are used in the building to operate the fan or for other purposes some of the sections should be so connected that exhaust steam can be used.

Sometimes the heaters are located at lower elevations than the steam boiler, or, at all events, at such low levels that water of condensation cannot be returned to the boilers. When such is the case, or when exhaust steam is used, either atmos-

135

pheric steam traps must be used in connection with the heaters, so that the water of condensation can discharge to waste, or return pumps or other apparatus must be provided to return the water of condensation to the boilers. Automatic air valves are generally used for keeping the coils free from air.

Cast-iron radiators may be used instead of pipe coils, but they are not so satisfactory, and are not extensively used.

Heat Emitted by Coils.—There is a vast difference between the amount of heat emitted by radiators and coils in still air, such as the air of rooms heated by direct radiation, and the amount of heat that will be emitted by the same surface of radiator or coils under forced blast from a fan. The reason for this is that the heat contained in the steam within the coils can be supplied much faster than it can be absorbed by the air surrounding the coils, no matter how fast, within practical limits, it is moved; it stands to reason, therefore, that the more air brought into contact with the coils the more heat will be carried off; consequently the more will be emitted by the steam. The actual heat emitted by coils, then, depends not only on the difference in temperature between the air and the steam, as in direct radiation, but, furthermore, on the velocity of air forced through the coils. In practice the quantity of heat given off per square foot of surface in a coil depends further on the depth or number of rows in the coil. The rate of emission is greatest when there is only one row of pipes in a coil, but the total quantity of heat emitted with one row of pipe is not sufficient for heating purposes, so that a number of rows must be provided, even though the average rate of emission is greatly reduced thereby. It has been found that the first section of the heater, four pipes deep in a forced-blast heater coil, emits about 40 per cent. of the total heat; the second section of four rows about 25 per cent.; the third about 15 per cent.; the fourth about 10 per cent.; the fifth about 6 per cent., and the sixth about 4 per cent. It will be observed, therefore, that after six sections of four rows each, or twenty-four rows of pipes, have been passed that little or no further heat is absorbed by the air. At all events, the amount is so small that twenty-four rows of pipes may be considered as the maximum depth of heater coil.

The amount of heat transmitted per hour at different velocities per square foot of surface for each degree difference between the air in contact with the coils and the steam within can be found in Table XIII.

It will be noticed that as the velocity of the air increases the amount of heat given off by the coils likewise increases, so that a one-row pipe coil which emits only 8.2 heat units per hour at an air velocity of 360 feet per minute will give off 19.5 heat units at an air velocity of 3,600 feet per minute. In Turkish bath ventilation advantage cannot be taken of that fact, however, for a velocity of 600 feet per minute through the coil is about all that should be calculated. At this velocity, with a coil sixteen rows deep and steam at a temperature of 300° Fahrenheit in zero weather, each square foot of heating surface would give off an average of $300 \times 6 = 1,800$ heat units per hour. This would be equivalent to a condensation of about 1¾ pounds of steam per square foot per hour.

Heaters made up of cast-iron indirect pin-type radiators will not give off more than one-half the amount of heat that will be emitted by an equal amount of surface of 1-inch pipe coil.

TABLE XIII.

HEAT EMITTED BY COILS TO AIR FROM FANS

Number of Rows of Pipe	Heat Units Emitted per Hour per Square Foot per Degree Difference Between the Temperature of Air in Contact with the Coils and the Steam Within							
	Velocity of Air in Feet per Minute							
	360	600	722	960	1,200	1,800	2,400	3,600
1	8.2	9.9	10.7	12.0	13.1	15.5	17.6	19.5
4	7.1	8.5	9.2	10.3	11.3	13.4	15.2	17.0
8	6.1	7.3	7.8	8.7	9.5	11.3	12.9	14.5
12	5.5	6.6	7.0	7.8	8.4	9.9	11.1	12.8
16	5.1	6.0	6.4	7.0	7.6	8.8	9.9	11.6
20	4.8	5.5	5.9	6.5	7.1	8.2	9.2	10.9
24	4.3	5.3	5.6	6.2	6.7	7.7	8.6	10.2

Steam Connections to Coils.—The size of steam and return pipes for heater connections not over 100 feet long, for both high and low-pressure steam can be found in Table XIV. In this table the amount of radiating surface is given both in lineal and in square feet. In computing the number of square feet in a coil of 1-inch pipe, 3 lineal feet of pipe were allowed for each square foot of surface. As a matter of fact there are 148.68

square inches in 3 lineal feet of wrought pipe, but that is close enough to 144 square inches for all practical purposes.

In proportioning coils to use exhaust steam from 75 to 100 feet of 1-inch pipe per horse power should be allowed, and the s u p p l y connection should be not less than 3 inches for engines up to 30 horse power; 4 inches for engines from 30 to 50 horse power; 5 inches for engines of 50 to 80 horse power; 6 inches for 80 to 120 horse power, and 8 inches for engines of 160 to 220 horse power.

TABLE XIV.
SIZE OF STEAM AND RETURN PIPE FOR COILS

Lineal Feet of 1-Inch Pipe in Heater or Coil	Square Feet of Pipe in Coil	High Pressure		Low Pressure	
		Diameter of Steam Pipe in Inches	Diameter of Return Pipe in Inches	Diameter of Steam Pipe in Inches	Diameter of Return Pipe in Inches
100 to 1,000	33 to 333	1½	1	3½	2½
1,000 to 2,000	333 to 666	2	1¼	4	3
2,000 to 3,000	666 to 1,000	2½	1¼	4½	3
3,000 to 4,000	1,000 to 1,333	2½	1½	5	3½
4,000 to 5,000	1,333 to 1,666	3	2	6	3½
5,000 to 6,000	1,666 to 2,000	3	2	7	4

Area of Heaters.—In proportioning the heater coil it is well to allow a free area between the pipes of about double the area of the fan outlet. There are two reasons for this. In the first place an allowance of about 50 per cent. greater area should be made on account of friction, if the same velocity were to be maintained through the heater as through the fan. But in Turkish bath work a slower velocity is desired, owing to the fact that air must be discharged through the system at an exceedingly slow speed, in order to avoid draughts. In the fresh air inlet and through the fan a velocity of about 900 feet per minute is satisfactory, but through the heater coil the air must be passed at a much slower rate, a velocity of about 600 feet per minute being about right. By increasing the free area of space between the coils, then, by about double the area of the fan discharge, or perhaps a little larger, would bring the velocity down to about the right limit. Even should the velocity through the coil drop slightly below 600 feet per minute there would be no perceptible loss of heat, for the rate of emission for a drop of speed of 25 feet would not average over .1 of a heat unit per square foot of surface.

Heat Required for Ventilation.—In ordinary heating and ventilation work sufficient heated air must be supplied to replace the heat lost by convection and radiation from the walls, the heat lost by leakage around doors, windows and other openings and that carried off through the ventilation flues. In Turkish bath work, however, the principal rooms are heated by direct radiation, so that all that will be required of the heater coils connected with the ventilation system is to heat the air forced into the rooms. Once the air reaches the rooms, instead of being cooled by loss through the walls it is maintained at its incoming temperature, if not actually raised in temperature by the steam coils within.

If the dressing rooms, sleeping rooms and all other compartments outside of the sweat rooms, shampoo room and plunge room are to be heated by direct radiation, then the only provision that need be made in the ventilation system is to heat enough air for ventilation purposes. If, on the other hand, the rooms outside of the active bathing rooms are to be heated as well as ventilated by the hot blast system, then allowance will have to be made to supply a sufficient volume of air at a sufficiently high temperature to heat the walls, floor, ceilings, cubical contents, and replace that heat lost by leakage, convection and radiation. Once the volume of air to be heated has been determined the size of coil required to heat that amount of air can be easily calculated. In doing so Table XV., giving the number of cubic feet of air warmed 1° by 1 heat unit, will be found convenient.

PROPERTIES OF AIR

Temperature of Air for Ventilation.—The air supply for ventilation must enter the several compartments at or above the temperature of the rooms, according to the system of ventilation adopted. If ordinary rooms are to be both heated and ventilated by the hot-blast system the air delivered to the rooms must have a temperature of about 120° in order to maintain the temperature of the compartments at 70° Fahrenheit. When the air is used for ventilation purposes only, on the other hand, it need not be heated any higher than the temperature of the compartment into which it is discharged. There are so many different temperatures to be maintained in the various compartments of a Turkish bath establishment that some

TABLE XV.—PROPERTIES OF AIR

1	2	3	4	5	6	7	8	9
Temperature of Air in Degrees Fahrenheit	Mixture of Air Saturated with Vapor — Ratio of Water to Dry Air	Ratio of Dry Air to Water Vapor	Cubic Feet of Vapor from 1 Pound of Water at Pressure Stated in Column 9	Heat Units Absorbed by 1 Cubic Foot of Dry Air per Degree Fahrenheit	Heat Units Absorbed by 1 Cubic Foot of Saturated Air per Degree Fahrenheit	Cubic Feet of Dry Air Warmed 1° Fahrenheit by One Heat Unit	Cubic Feet of Saturated Air Warmed 1° per Heat Unit	Elastic Face of Vapor in Inches of Mercury (Regnault)
0	.00092	1,092.4		.02056	.02054	48.5	48.7	.044
12	.00115	646.1		.02004	.02006	50.1	50.0	.074
22	.00245	406.4		.01961	.01963	51.1	51.0	.118
32	.00379	263.81	3,289	.01882	.01924	52.1	51.8	.181
42	.00561	178.18	2,252	.01847	.01884	53.2	52.8	.267
52	.00819	122.17	1,595	.01818	.01848	54.0	53.8	.388
60	.01251	92.27	1,227	.01811	.01822	55.0	54.9	.522
62	.01179	84.79	1,135	.01777	.01812	56.2	55.7	.556
70	.01680	64.59	882	.01744	.01794	57.3	56.5	.764
72	.01780	59.54	819	.01710	.01790	58.5	56.7	.785
82	.02361	42.35	600	.01690	.01770	57.2	56.8	1.092
92	.03289	30.40	444	.01682	.01751	58.5	57.1	1.501
100	.04495	23.66	356	.01651	.01735	59.1	57.8	1.929
102	.04547	21.98	334	.01623	.01731	59.5	57.8	2.086
112	.06253	15.99	253	.01596	.01711	60.6	58.5	2.731
122	.08584	11.65	194	.01571	.01691	61.7	59.1	3.621
132	.11771	8.49	151	.01544	.01670	62.5	59.9	4.752
142	.16170	6.18	118	.01518	.01652	63.7	60.6	6.165
152	.22465	4.45	93.8	.01494	.01654	65.0	60.6	7.930
162	.31713	3.15	74.5	.01471	.01656	66.0	60.4	10.099
172	.46338	2.16	59.2	.01449	.01658	67.1	60.3	12.758
182	.71300	1.402	48.6	.01466	.01687	68.0	59.5	15.960
192	1.22643	.815	39.8	.01406		68.9		19.828
202	2.80230	.357	32.7			68.5		24.458
212	Infinite	.000	27.1			71.4		29.921

For ordinary calculations it may be assumed that 1 heat unit will raise the temperature of 60 cubic feet of air 1° Fahrenheit.

method must be adopted either to temper the heated air going to some rooms so it will not be too warm or to reheat the air for other rooms so it will not be too cold. The size of coil to use in connection with the fan cannot well be determined until this matter is decided upon. If the intention is to heat all air to the temperature of the hottest sweat room and temper with cold air that going to the cooler compartments, then a quantity of heat equal to that required to warm the air used for ventilation in the several rooms will have to be supplied by the coil. If, on the other hand, reheaters will be used, the air for ventilation will have to be raised by the main coil only to about 75° or 80° Fahrenheit, depending on how far it will have to travel and the probable loss of heat before reaching the several room. In that case the extra heat required for the sweat and the other bath rooms will have to be supplied by auxiliary heaters, as will be explained later.

It might be inferred from what has already been said in another chapter that it is better to use reheaters to raise the temperature of the air to the right degree than to raise the temperature of all the air to a very high degree, then temper it with cool air before discharging it into the various rooms. Very hot air seems to be burnt. It is possible that as water is an almost universal solvent for minerals and liquids, air is likewise an almost universal solvent for gases; and that just as potable waters often carry beneficial foreign matters in solution, so air carries other gases which in their natural condition are beneficial to the system. But these voletile properties of the air are very elusive, and a high temperature is liable to burn them out or drive them off, thus depriving the air for ventilation of some of its most valuable qualities or properties.

There is no denying the fact that no matter how much air we heat and force into a building in form of ventilation, the air of the interior of the building never approximates in wholesomeness and invigorating qualities of the air outside, particularly the crisp air of a Spring or Autumn morning.

CHAPTER X

AIR-DISTRIBUTING SYSTEMS

ꆛ

SINGLE-DUCT System.—There are two systems of air distribution commonly used in heating and ventilating work, which are known, respectively, as the single-duct system and the double-duct system. The single-duct system is shown in Fig. 31 and would be used in Turkish-bath buildings where the sleeping, dressing and other rooms, outside of those devoted to the bath proper, would be heated and ventilated by the hot-blast system. In such cases the air for heating and ventilating these rooms would be raised to a temperature not exceeding 140° Fahrenheit, and more probably not to a greater temperature than 120° Fahrenheit, the exact temperature depending on how far the air had to be forced through ducts and flues, the amount of heat it would lose in passage and the excess heat required to replace that lost by radiation and convection from the walls of the rooms. Such temperatures would not be sufficiently high for the sweat rooms of the bath, however, so that in order to raise the temperature of the air to the right degree for those compartments indirect stacks, or re-heaters, would have to be provided at the base of the flues, as indicated in the illustration. Sufficient surface is provided in these stacks to raise the temperature of the air for the several rooms to the required temperatures. Cast-iron indirect radiators are shown in this illustration, although pipe coils of suitable design, or any other kind of heating surface, may be used.

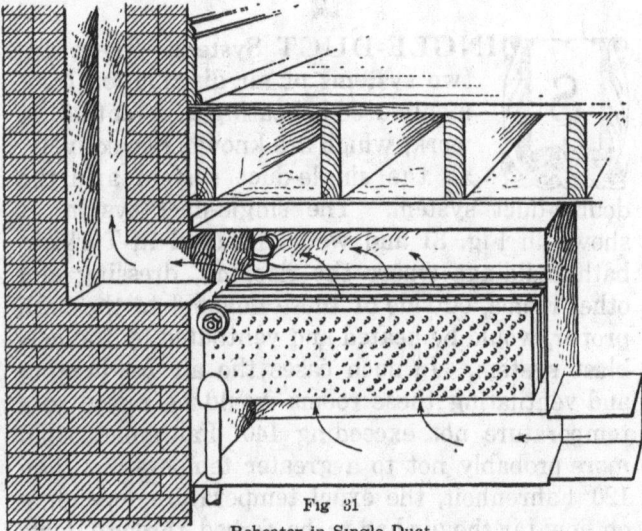

Fig 31
Reheater in Single Duct

Double-Duct System.—The application of a double-duct system of ventilation is shown in Fig. 32. Two ducts are run to each flue, one duct above the other, the under duct usually being the cold one. At each vertical flue a mixing damper is provided so that cold air, hot air, or air of any temperature between the two extremes can be had

146

in the rooms by simply manipulating the dampers. Such a system would be necessary where the air for the sweat rooms was heated to the required temperature for the hottest of these compartments, so that air for the other rooms could be tempered with cold air to the several temperatures required in the various rooms

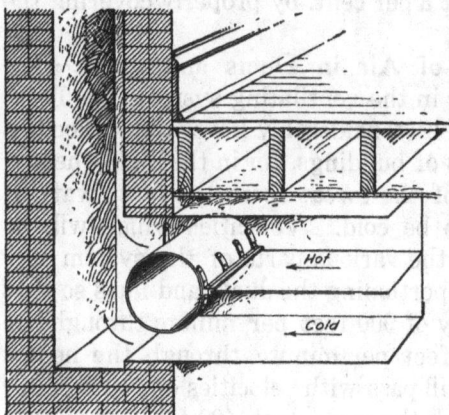

Fig. 32
Double Duct

The ducts for ventilation purposes, whether of the single or double-duct system, may be run as a single main or in a divided main with branches to the various flues, or separate ducts may be run from the plenum chamber to the several flues.

Square or rectangular ducts are generally used in preference to round ones, notwithstanding the fact that round ducts are cheaper, stronger and offer less frictional resistance to the flow of air. The ducts should be covered with several thicknesses of asbestos paper, corrugated asbestos air-cell covering, or some equally good heat-insulating material, and in exposed places the covering can be finished by drawing canvas tight over the covering and painting the canvas to harmonize with the surroundings. The loss of heat from uncovered ducts, with velocities ordinarily used in practice, will average about 7 per cent. per 100 lineal feet, while with the slow velocities required for Turkish-bath work will average perhaps 10 per cent. per 100 lineal feet. This heat loss can be reduced to about 3 per cent. by properly covering the ducts.

Velocity of Air in Ducts and Flues.—The velocity of air in the ventilating system of a Turkish bath must be considerably lower than velocities in other types of buildings, for in the highly heated temperature of the sweat room the least waft of air is felt to be cold. Velocities which will be found safe in the various parts of the system may be had by proportioning the ducts and flues so that with a velocity of 900 feet per minute through the fan and 600 feet per minute through the heater coils the air will pass with velocities of 500 feet per minute through the main duct, 400 feet per minute through the branches, 300 feet per minute through the vertical stacks and enter the hot rooms at velocities not above 200 feet per minute. For the other parts of the building much higher velocities would be permissible, even as high as 900 to 1,200 feet per minute in the main ducts, 700 to 1,000 feet per minute in the branches, 500 to 800 feet in the vertical flues, and 300 to 400 feet through the registers. As the highest velocity available, however, is 600 feet per minute through the heater, this part of the system can be proportioned for a velocity of only 600 feet per minute through the main duct, 500 feet per minute through the branches, 400 in the vertical stacks and 300 through the registers. The air to other parts of the building could be moved with the same velocities as for the hot rooms, but the faster the air is moved the less is the percentage of loss of heat from the ducts and the smaller the size of flues and ducts required.

Location of Inlet and Outlet Registers.— The location of inlet and outlet registers is of the utmost importance in ventilation, for if they are improperly placed the system will be a failure, no matter how well proportioned and installed the other

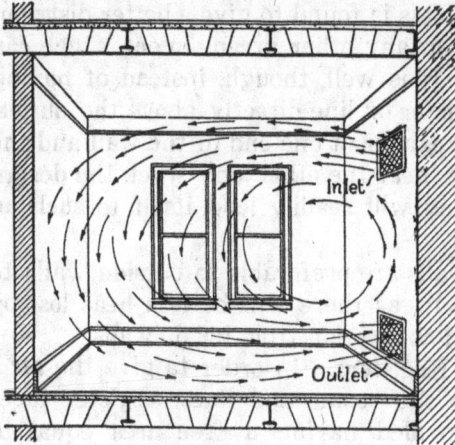

Fig. 33
Inlet and Outlet Registers

parts of the work may be. The best location for inlet and outlet registers is shown in Fig. 33. The inlet register is located the higher of the two, and should be at least 8 feet above the floor level; on the other hand, as the object of the outlet register is to remove continuously the bottom layer of air it should be placed at the floor level, and in the

149

hot rooms by the floor level is meant actually so, not several inches above the floor level. In the other rooms of the compartment a few inches more or less do not matter so much. In the hot rooms long, narrow registers, 6 or 8 inches deep, are preferable to narrower but higher designs. The coldest air in the rooms is right at the floor line, and the coldest air is what should be removed. In addition to the respective elevations of the two registers it will be noticed they are placed in the same wall. This is found to give a better distribution of air than any other arrangement which can be devised. It is well, though, instead of having the inlet register on line directly above the outlet, to have one register at one end of the wall and the other register near the other end when the design of the building will readily lend itself to such an arrangement.

Inside walls are preferable to outside walls to run the flues in, as there will be less heat lost by radiation and convection from inside walls.

Size of Registers.—In order to give the air a fair inlet to the rooms and flues, register faces must be provided having a free area equal in capacity to that of the flues. It is found by experiment and in practice that the resistance offered by ordinary commercial registers decreases their actual capacities to such an extent that the area of openings must exceed the area of the flues by 33 per cent. In the hot rooms of a Turkish bath, where the velocities are low, in order to interpose as little resistance as possible large openings in the register face will be found preferable to small openings, and the area of the openings may be with benefit 50 per cent. greater than the area of the flues.

REGULATION OF TEMPERATURE

Hand Regulation of Temperature.—All the heating coils used in connection with the Turkish bath should be installed in sections, so that one, several, or all can be cut out or turned on as occasion requires to maintain the right temperature in the several rooms. Each section must be suitably valved for this purpose, and, as there are numerous attendants keeping close watch on temperature and other conditions conducive to the welfare of the bathers, the regulation of temperature may well be left to them, and ordinary steam valves only installed to control the several sections of coils.

Automatic Regulation of Temperature.— If the amount of money available for building and equipping the bath will warrant the extra outlay the temperature of the several rooms may be maintained automatically by means of a Johnson or other type of temperature regulator. The Johnson regulator consists of a thermometer or thermostat placed in the room where the temperature is to be regulated, an air compressor and air tank in the basement or other convenient part of the building and a diaphragm valve operated by compressed air on the several steam coils. The temperature is controlled by the difference in expansion between two strips of different metals. Expansion of the metals opens a small air valve which admits compressed air to the top of a diaphragm valve, thus closing the valve and shutting off steam from that coil. Contraction of the strips of metal closes the air valve, relieves the air from the diaphragm valve and permits that valve to open, thus turning steam into the coil. Thermostatic temperature controllers will operate within a few degrees of temperature, so that the air of a room can be maintained practically at a constant temperature.

An apparatus for automatically controlling the supply of hot air to rooms heated by the hot-blast system of heating and ventilation is shown in Fig. 34. This is operated in much the same manner as

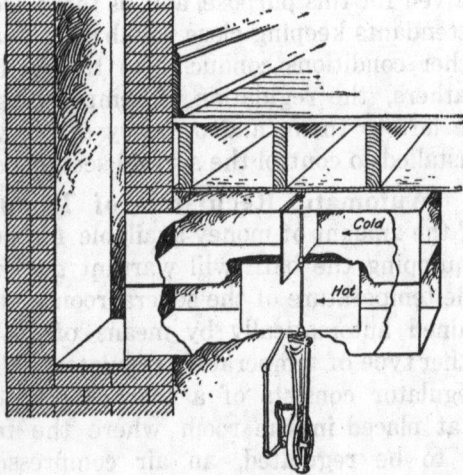

Fig. 34
Automatic Regulator

the diaphragm steam valves are regulated, the only difference being that a motor is used to operate the dampers, whereas a diaphragm valve is used to turn on and shut off steam from the steam coils.

A motor can be had which will shut off completely the supply of hot air and turn on the cold or tempered air whenever the temperature of the air within the room warrants such a change, or a motor can be provided which will gradually open and close the dampers or hold them at such angles that the right proportions of hot and cold air will be delivered at all times.

AIR MOISTURE FOR TURKISH BATHS

Outside of the hot rooms, where the air is preferably dry, the other compartments of a Turkish bath should be provided with sufficient moisture to maintain the normal humidity of the atmosphere. The normal humidity is about 60 to 70 per cent. of saturation of the relative humidity. By relative humidity is meant the percentage of moisture contained in air at different temperatures. For example, at 10° Fahrenheit and 60 per cent. of saturation 1 cubic foot of air would contain .466 grains of water, while at a temperature of 70° Fahrenheit 1 cubic foot of air would contain 4.788 grains of moisture when containing 60 per cent. of saturation. The amount of water, it will be seen, varies thus with the temperature for certain percentages of saturation, and the amounts so contained at different temperatures, but the same percentage of saturation represents the relative humidity at different temperatures. The actual amount of moisture that air will hold depends upon its temperature. When the air is so charged that it will hold no more it is saturated, or contains 100 per cent. at that temperature. Adding more moisture to the air or lowering its temperature will cause a precipitation or condensation—as, for instance, in rain or dew, which are merely a precipitation of moisture due to over saturation or fall in temperature. On the other hand, if air is saturated and its temperature is then raised the percentage of saturation falls and the air can absorb more moisture. The relative humidity of air at different temperatures can be found in Table XVI.

TABLE XVI.—RELATIVE HUMIDITY OF AIR

| Temperature Degree | Weight of Water in Grains per Cubic Foot of Air | | | | | | |
| | Percentage of Saturation | | | | | | |
	40	50	60	70	80	90	100
—10	.114	.142	.171	.200	.228	.256	285
— 5	.148	.185	.222	.259	.296	.333	.870
0	.192	.240	.289	.337	.385	.433	.481
+ 5	.244	.305	.366	.427	.488	.549	.610
10	.310	.388	.466	.543	.621	.698	.776
15	.394	.493	.592	.690	.789	.887	.986
20	.494	.618	.741	.864	.988	1.112	1 235
25	.620	.776	.931	1 086	1 241	1.396	1.551
30	.774	.968	1.161	1 354	1.548	1.742	1.935
35	.946	1.183	1.420	1 656	1 893	2 129	2 366
40	1.140	1.424	1.709	1 994	2 279	2 564	2 849
45	1.366	1.707	2.048	2.390	2 731	3.073	3.414
50	1.630	2 038	2 446	2 853	3 2n1	3.668	4 076
55	1 940	2 424	2.909	3 394	3 879	4.364	4.849
60	2.298	2.872	3.447	4 022	4.596	5 170	5 745
65	2.713	3 391	4.069	4.747	5 426	6.104	6 782
70	3 192	3 990	4.788	5 586	6.384	7.182	7 980
75	3.742	4 678	5.614	6 549	7.485	8 420	9 356
80	4.374	5.467	6.560	7.654	8.747	9.841	10 934
85	5.094	6.368	7.642	8.915	10 189	11 462	12.736
90	5 916	7.395	8 874	10.353	11.832	13 311	14 790
95	6 850	8 562	10.274	11 987	13 699	15 412	17 124
100	7.906	9.883	11.860	13 836	15.813	17.789	19 766
105	9.100	11 375	13.650	15 925	18 200	20 475	22 750
110	10.445	13.056	15 967	18 278	20 890	23.501	26 112
115	11.955	14.943	17.932	20 921	23 910	26 898	29 887
120	13.646	17 057	20.468	23.880	27.880	30 703	34 115

From the table it will be seen that if air is introduced to a building during zero weather when the humidity of the outside air is normal or contains about 60 per cent. of moisture, 1 cubic foot of air at that temperature would contain only

.289, or slightly over one-quarter grain of water. When this cubic foot of air is raised in temperature to 70° Fahrenheit, however, it will require 4.788, or over four and three-quarters grains of moisture, to raise it to the relative and normal humidity of 60-per-cent. saturation.

This moisture must be supplied to the air in some manner or a feeling of discomfort will ensue

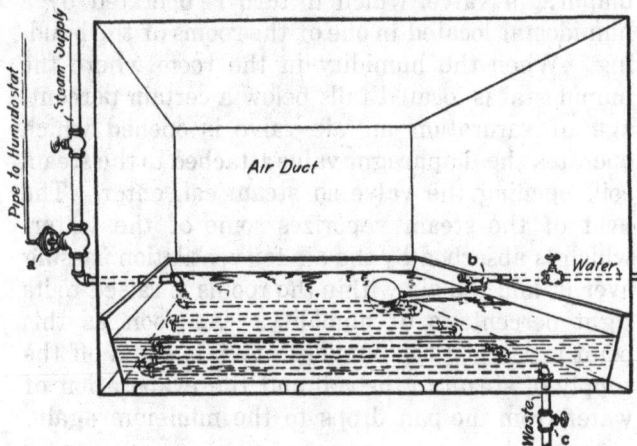

Fig. 35

Humidostat

and moisture will be absorbed from the woodwork of interior trim, furniture, or from whatever other source it is available. In large bathing establishments, where a fountain is provided in the resting room, sufficient moisture will be supplied from this source, which at the same time can be made an ornament of the room. When there is no fountain, on the other hand, a humidostat should be provided.

Humidostats.—A humidifier for moistening air is shown in Fig. 35. It consists of a shallow pan placed in a hot air duct and kept filled with water through a ball cock supplied from the city water. A steam coil is immersed in the water and the supply of steam to the coil is controlled by a diaphragm valve, which in turn is operated by a humidostat located in one of the rooms of the building. When the humidity in the room where the humidostat is located falls below a certain percentage of saturation an air valve is opened which operates the diaphragm valve attached to the steam coil, opening the valve so steam can enter. The heat of the steam vaporizes some of the water, which is absorbed by the air for ventilation passing over it, and the air within the rooms is raised to its right percentage of humidity. As soon as this point is reached the humidostat again shuts off the supply of steam to the coil and the evaporation of water from the pan drops to the minimum again.

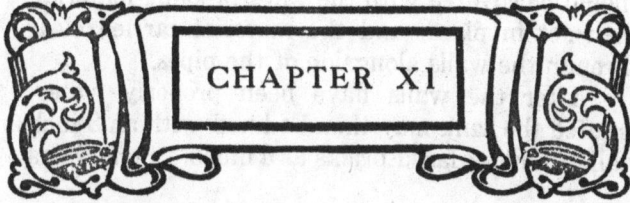

CHAPTER XI

TURKISH BATH DETAILS

PLUNGE BATHS

CONSTRUCTION of Plunge Baths.— The general construction of a plunge bath depends somewhat upon the place where it will be located. When the pool will be located in the basement, first floor, or in any other place where it will rest upon the ground, it may be made as shown in Fig. 36. A concrete tank with walls 10 inches or more thick, according to size and depth, reinforced with expanded metal, is waterproofed by applying a coating of asphalt, hydroline, hydrex, biolite, or other good waterproofing material about ⅜-inch thick, laid hot in burlap of 8 to 12-ounce weight, then while still hot covered with another ⅜-inch coating of the waterproofing material and burlap course, and pressed and broomed to a perfect bond. After the tank has been waterproofed it may be tested by filling with water and allowing to stand for several days to see if leaks are disclosed. Where pipes pass through the walls the pipes should be covered with the waterproofing material, then set in place, and the material carried clear through the walls alongside of the pipes.

After the walls have been properly waterproofed the tank may then be lined with an 8-inch wall of white glazed bricks as a finish. Sometimes

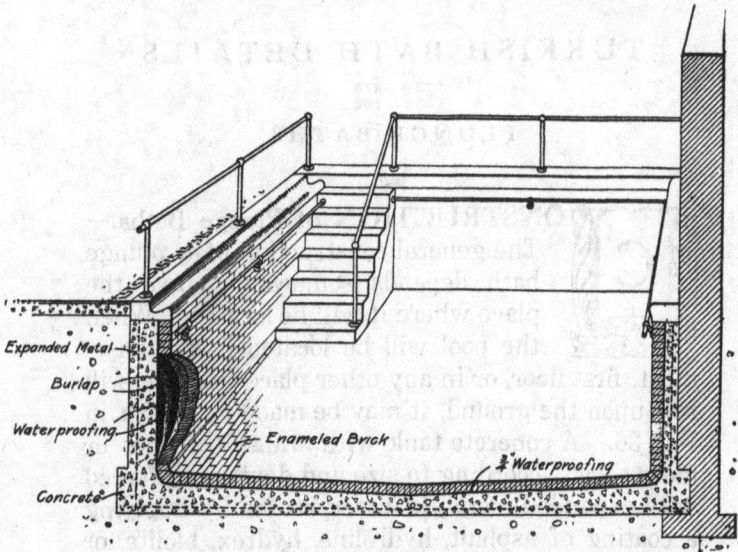

Fig. 36

Plunge Tank

only 4-inch walls of brick are used, but in such cases, if there is danger in back pressure from ground water, the site will have to be underdrained. Instead of using bricks, the waterproof course may be laid between two walls of concrete, and the inner surface of the tank covered with

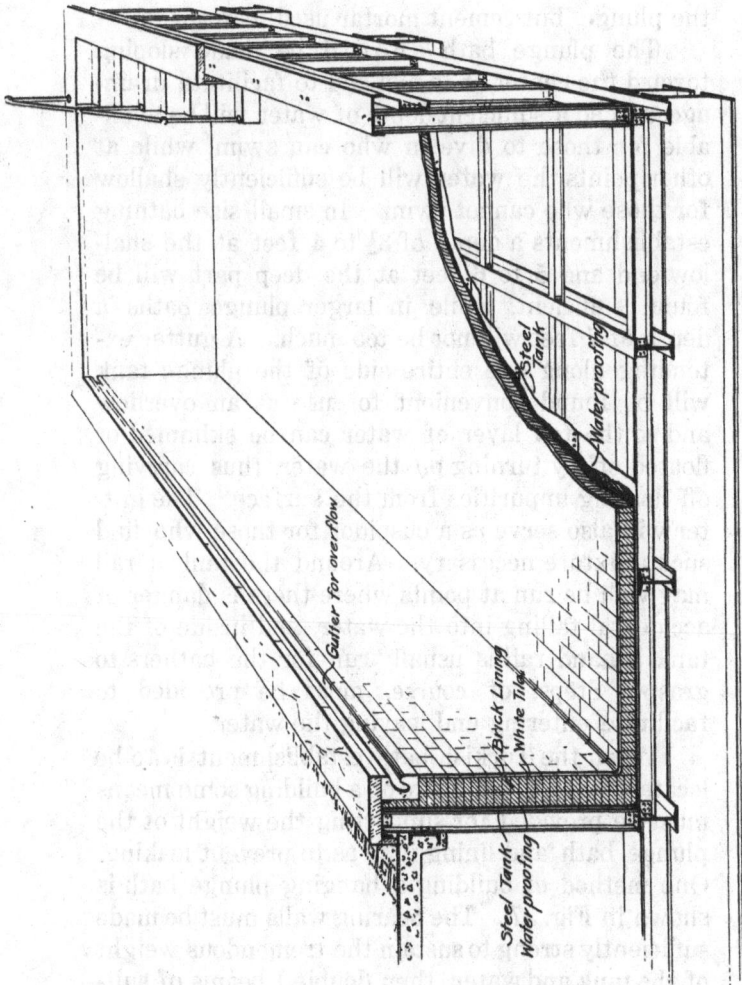

Steel
Tank

Waterproofing

Gutter for overflow

Brick Lining
White Tile

Steel Tank
Waterproofing

Fig. 37

A Hanging Plunge Bath

159

glazed tile. No lime mortar should be used around the plunge, but cement mortar used instead.

The plunge bath should be made sloping toward the center or to one end to facilitate drainage, and so a sufficient depth of water will be available for those to dive in who can swim, while at other points the water will be sufficiently shallow for those who cannot swim. In small size bathing establishments a depth of 3½ to 4 feet at the shallow end and 5 to 6 feet at the deep part will be found sufficient, while in larger plunge baths a depth of 7 feet will not be too much. A gutter extending along one entire side of the plunge tank will be found convenient for use as an overflow and so the top layer of water can be skimmed or floated off by turning on the water, thus carrying off floating impurities from the surface. The gutter will also serve as a cuspidor for those who find such a fixture necessary. Around the tank a rail may well be run at points where there is danger of accidental falling into the water, and inside of the tank a hand rail is usually run for the bathers to grasp. Steps, of course, must be provided to facilitate entering and leaving the water.

When the Turkish-bath establishment is to be located in the upper floors of a building some means must be provided for supporting the weight of the plunge bath and lining it so as to prevent leaking. One method of building a hanging plunge bath is shown in Fig. 37. The bearing walls must be made sufficiently strong to sustain the tremendous weight of the tank and water, then double I beams of suitable strength will be required at frequent intervals to support the framework for the tank. The framework will be made of I beams well riveted together so as to form base and side supports, and inside of this framework a steel tank must be built as a casing for the plunge. This steel tank must be

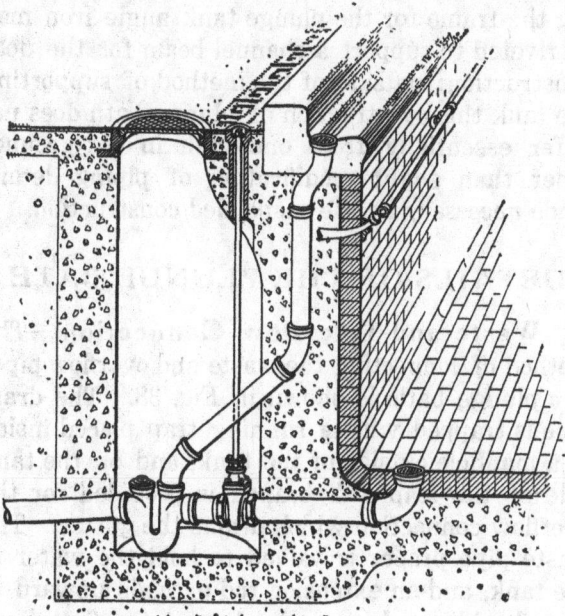

Fig. 38

Waste Connection to Plunge Tank

made of sufficiently heavy plates to sustain the weight of lining and water between supports without bending or giving, and should be well coated on the outside with some good metallic paint to prevent corrosion. Inside the tank should be

161

waterproofed with a couple of good coats of waterproofing and burlap as previously explained, then
if perfectly tight may be concreted and lined with
tile, glass or marble, or may be finished with an 8-
inch lining of glazed bricks, laid up in cement mortar. On the outside of the upright I beams forming the frame for the plunge tank angle iron may
be riveted to support a channel beam for the floor
construction; outside of the method of supporting
the tank this construction of plunge bath does not
differ essentially from one built in the ground,
other than slight modification of piping details
made necessary by the suspended construction.

DETAILS OF THE PLUNGE BATH

Waste and Overflow Connection.—The
method of connecting the waste and overflow pipes
to a plunge bath is shown in Fig. 38. The drain
pipe is trapped with a running trap placed inside
of a manhole outside of the tank, and on the tank
side of the trap a Y branch is provided for the
overflow connection which drains the gutter. The
waste pipe proper is valved to hold the water in
the tank, and an extension rod extends upward to
the wheel located in a valve box at the floor level
so the valve can be easily opened or closed. All
pipe, where bedded in the masonry of the tank,
should be coated with the waterproofing material
in order to form a water-tight joint or bond at
these points.

The overflow pipe should be made large enough
to carry off as much water as can be discharged

into the tank by the supply pipe so there will be no danger of the tank overflowing, and the waste pipe should be made sufficiently large so that the tank can be emptied in an hour or less. The size of drain pipe capable of emptying the tank in this length of time can be found in Table XVII.

The quantities given in the table are in cubic feet per minute, so to find the discharging capacities of the pipes per hour the quantities must be multiplied by 60, as there are 60 minutes in an hour. Instead of doing this the cubical contents of the plunge bath may be divided by 60, which will give the rate at which it must be emptied per minute, then a pipe of that capacity can be selected from the table. The quantities in the table are in cubic feet. These, however, can be converted into gallons by multiplying them by 7.5, which is approximately the number of gallons in a cubic foot.

Sometimes plunge tanks are located below the level of the street sewer, in which case the contents of the tank must be elevated to the sewers by means of pumps or compressed air ejectors. Ordinarily a centrifugal pump direct connected to an electric motor and located in a sump adjoining the plunge bath or at some other convenient point nearby will be found the best method for raising the water.

Hand-rail Supports.—One method of providing anchors for the support of hand rails is shown in Fig. 39. A piece of wrought pipe flattened on one end to keep from turning, and threaded and bent to the right angle, is built into

TABLE XVII.—SIZE OF DRAIN PIPES

Diameter	2 Inches		2½ Inches		3 Inches		4 Inches		5 Inches		6 Inches	
Fall Ft. in Ft.	Velocity Feet per Minute	Discharge Cubic Feet per Minute	Velocity Feet per Minute	Discharge Cubic Feet per Minute	Velocity Feet per Minute	Discharge Cubic Feet per Minute	Velocity Feet per Minute	Discharge Cubic Feet per Minute	Velocity Feet per Minute	Discharge Cubic Feet per Minute	Velocity Feet per Minute	Discharge Cubic Feet per Minute
1 in 20	273	5.46	297	8.91	335	13.40	390	32.40	432	58.32	480	93.60
1 in 25	246	4.92	273	8.19	300	12.00	345	28.64	397	52.25	450	87.75
1 in 30	220	4.40	249	7.49	270	10.80	312	25.89	351	47.39	390	77.65
1 in 35	204	4.08	228	6.84	250	10.00	288	23.80	324	43.74	360	70.20
1 in 40	192	3.84	216	6.48	237	9.48	272	22.68	306	41.31	330	64.35
1 in 45	180	3.60	201	6.03	222	8.88	255	21.16	288	38.88	315	61.42
1 in 50	174	3.48	192	5.76	210	8.40	243	20.17	272	36.72	300	58.50
1 in 60	153	3.06	174	5.22	190	7.60	216	17.93	245	32.07	270	52.65
1 in 70	144	2.88	162	4.86	177	7.08	204	16.98	229	30.91	252	49.14
1 in 80	135	2.70	150	4.50	165	6.60	198	16.43	210	28.35	214	45.63
1 in 90	129	2.50	144	4.32	156	6.24	180	14.94	201	27.13	222	42.29
1 in 100	120	2.40	135	4.05	150	6.00	170	14.11	192	25.92	210	41.16

Diameter	7 Inches		8 Inches		9 Inches		10 Inches		11 Inches		12 Inches	
Fall Ft. in Ft.	Velocity Feet per Minute	Discharge Cubic Feet per Minute	Velocity Feet per Minute	Discharge Cubic Feet per Minute	Velocity Feet per Minute	Discharge Cubic Feet per Minute	Velocity Feet per Minute	Discharge Cubic Feet per Minute	Velocity Feet per Minute	Discharge Cubic Feet per Minute	Velocity Feet per Minute	Discharge Cubic Feet per Minute
1 in 20	510	135.15	540	189	573	252	620	335	620	455	750	595
1 in 25	480	127.20	480	168	510	224	540	292	570	376	600	468
1 in 30	438	116.07	450	158	471	207	510	275	520	343	540	420
1 in 35	390	103.35	408	143	441	194	456	246	480	316	510	397
1 in 40	363	96.19	390	137	411	180	432	233	450	297	480	474
1 in 45	342	90.63	360	126	390	172	405	218	430	283	450	351
1 in 50	327	86.65	345	120	363	160	390	210	410	270	420	327
1 in 60	288	76.32	289	108	330	145	345	186	360	238	390	304
1 in 70	270	71.55	280	98	306	135	324	175	340	224	360	280
1 in 80	252	66.78	270	94	294	123	309	167	325	214	330	257
1 in 90	240	63.60	258	90	273	120	285	154	300	198	315	245
1 in 100	221	58.56	255	86	258	114	270	146	288	190	300	234

NOTE—To determine discharge in U. S. gallons multiply cubic feet by 7.5.

164

the wall. Then when the glazed bricks or tile, as the case may be, are in place the hand rail can be finished by putting the cast-brass skew plates on the anchor pipes close to the wall, slipping short pieces of brass tubing over the pipe, as shown in the illustration, then screwing into place the cast-brass ring heads. Instead of using iron pipe to bed in the walls, pieces of iron bar may be used, and either threaded for the ring head as in the

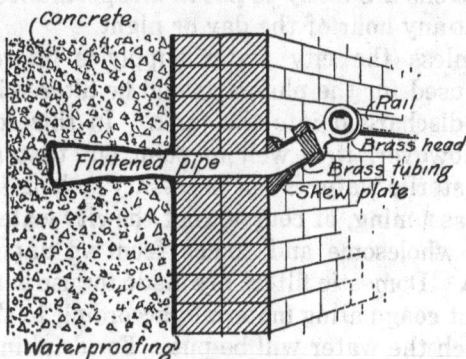

Fig. 39

Hand-Rail Support

case of iron pipe, or drilled and tapped for bolting the ring head to the iron bar. Copper or brass pipe may likewise be used instead of the wrought pipe and will be found to last longer.

Water Supply for Plunge Bath.—The water-supply pipe for the plunge bath should be made sufficiently large so that the tank can be filled in at least one hour's time. This necessitates the tank being out of use about two and one-half hours, allowing one hour for emptying, one hour for filling and thirty minutes for cleaning. To facilitate cleaning, a hose connection about two and one-half inches in diameter, located close to the tank, will be found convenient. Small tanks in medium-sized institutions can be emptied and filled in less than two and one-half hours, but in the larger institutions it is not advisable to have the plunge bath out of use for a greater length of time, for patrons are likely to put in an appearance for a bath at any hour of the day or night.

Unless the city water supply is filtered all water used for the plunge should be filtered before being discharged into the tank. In this connection, however, it is well to point out that a perfectly sterile water is not necessary so long as it is clear, assuming, of course, that the city water supply is wholesome and comes from an unpolluted source. Domestic filters are made for use with and without coagulating materials, according to the use to which the water will be put. For drinking and household use the filters are designed to not only clear the water, but render it sterile as well, by removing all bacteria. In industrial establishments and other places where the water is not for drinking purposes, or will be boiled before use— as, for instance, in canning works—all that is necessary is to clarify the water of all organic impurities. Where the water is to be made sterile, coagulant and coagulating apparatus must be used in connection with the filter, but for Turkish-bath work, outside of the drinking fountains, a clear water is all that will be necessary. Two filters,

166

therefore, may well be supplied where the water is not already filtered—one large filter, or a battery of two large filters, to supply the bath house proper, and a small filter with coagulating apparatus to sterilize the water for drinking fountains.

Heating Water for Plunge Baths.—Ordinarily in moderate climates there will be no necessity for heating the water in the plunge baths. The object of the plunge is to close the pores and administer the tonic effect of a cold bath, and both these functions will be better performed the colder the water in the tank, within reasonable limits. A temperature of 65° Fahrenheit is not too cold for the water in the plunge of a Turkish bath, although for an ordinary swimming pool, such as is sometimes built as an adjunct to a Turkish bath, a temperature of 75° Fahrenheit is more suitable.

There are a number of ways in which water may be heated, but very few of them are applicable to Turkish-bath work. Heating the water by injecting steam into the tank is objectionable for the reason that the process is more or less noisy. Heating by means of submerged coils in the plunge tank is objectionable for the reasons that the pipes are not only unsightly and in the way but, further, unless made of brass or copper, will rust and at all times prove to be dirt catchers. Within a tank where so many people bathe no nooks or crannies should be permitted which will catch, hold or conceal filth. Submerged coils, therefore, may be considered too objectionable for use in the plunge tank.

Steam or fire may be used for heating the water in a tank, but whichever kind of heat is used it should be so applied that water from the tank will have to circulate through the heater to be warmed.

Heating with Hot-water Heater.—When the installation is such that the heater can be

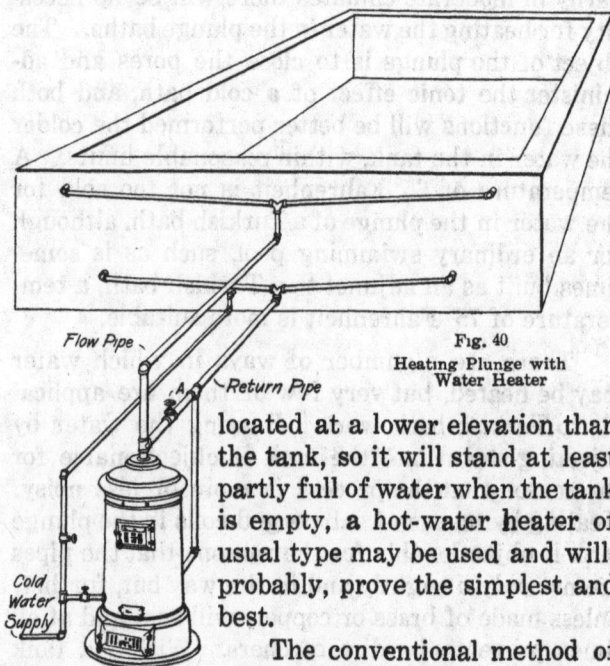

Flow Pipe

A

Return Pipe

Cold
Water
Supply

Fig. 40

Heating Plunge with
Water Heater

located at a lower elevation than the tank, so it will stand at least partly full of water when the tank is empty, a hot-water heater of usual type may be used and will, probably, prove the simplest and best.

The conventional method of connecting up a water heater to warm the water in a plunge bath is shown in Fig. 40. To use this kind of heater it is necessary that it be located as low down as the bottom of the tank, and preferably lower. A branch of cold-water

168

supply is then connected to the heater and the main supply continued on to the tank so that water can be supplied direct, or through the heater. Under no consideration should a valve be placed in either the flow or return pipe between the tank and heater. In small tanks only one inlet and outlet connection will be necessary, while in larger tanks multiple connections are advisable. The flow and return connections to the tank may be made at one side, as shown in the illustration, or in extremely large tanks the flow pipes may be connected to one side and the return pipes to the opposite side of the tank. With this type of heater to fill the tank the cold water can be turned on to the heater, the valve *a* opened and a fire lighted, thereby heating the water as it flows through. After the tank is full the cold-water valve can be shut off, the valve *a* between the flow and return pipe closed and the water in the tank circulated through the heater by gravity until the right temperature has been reached. If the bi-pass controlled by the valve *a* were omitted hot water would not flow out of the heater when the tank was filling, the cold water would short circuit right through the return pipe to the tank, while steam would blow through the flow pipe. Placing a valve or check in the return pipe would overcome this, but, as previously stated, checks or valves of any description are objectionable in the flow and return pipes.

Heating Water with Feed-water Heaters. —When the plunge bath or swimming pool is situated at a lower level than the heater the water cannot be circulated by gravity, but must be pumped through the heater. Instead of using a heater in such cases the water could be heated by discharging steam into it, but, as before stated, that practice is noisy, and there is more or less odor from the steam which would be objectionable in the plunge room of a Turkish bath.

One method of heating water at a lower level is shown in Fig. 41. In this method an ordinary feed-water heater, heated by a steam coil, is used, and the water circulated through it by means of a pump. When filling the tank the water can be passed through the feed-water heater, using live steam; then, when the tank is filled, the water can be circulated through the feed-water heater by means of the pump, and the exhaust steam from the pump used for heating the water. It will be observed that the feed water is connected with both the live steam pipe from the boiler and the exhaust steam from the pump, so that either may be used, and the pipes are all properly valved with this object in view. In the exhaust pipe to the roof there is a back pressure valve, so that in case the steam consumption in the pump exceeds that of the feed-water heater the excess can escape to the atmosphere without imposing an unnecessary back pressure on the pump.

The illustration shows the pump connected with a bi-pass to sewer, so the water can be discharged to waste instead of circulating it through the heater in case it is desired to empty the tank. This bi-pass will not be necessary, however, if the

170

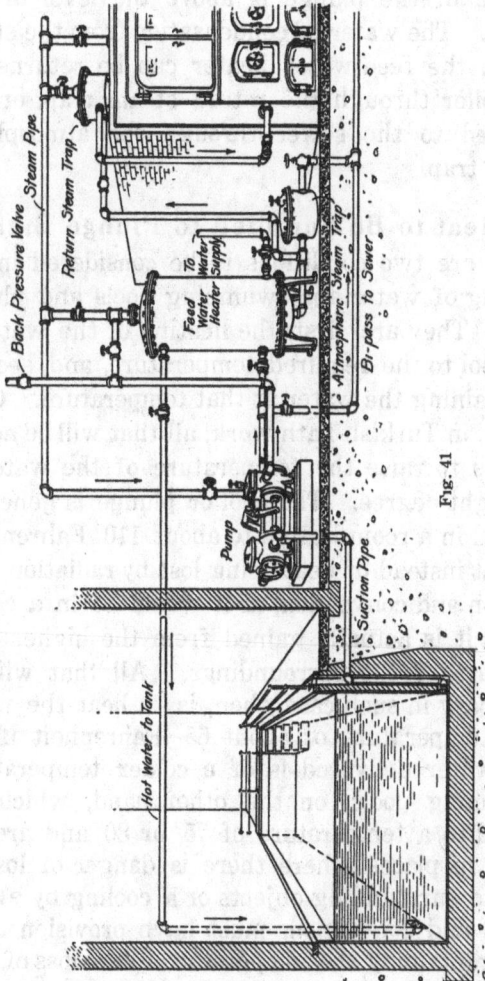

Fig. 41

Heating Plunge with Feed-water Heater

171

bottom of the plunge is above the level of the sewer. The water of condensation from the steam coil in the feed-water heater can be returned to the boiler through the return steam trap, or discharged to the sewer through the atmospheric steam trap.

Heat to Be Supplied to Plunge Baths.— There are two conditions to be considered in the heating of water for swimming pools and plunge bath. They are, first, the heating of the water in the pool to the required temperature, and, second, maintaining the water at that temperature. Ordinarily, in Turkish-bath work, all that will be necessary is to raise the temperature of the water to the right degree. The pool or plunge is generally located in a room heated to about 110° Fahrenheit, so that instead of heat being lost by radiation, conduction and convection, as it would be in a colder place, it is actually gained from the higher temperature of the surroundings. All that will be necessary in such case, then, is to heat the water to a temperature of about 65° Fahrenheit if the city water delivered is of a colder temperature. Swimming pools, on the other hand, which are heated to a temperature of 75° or 80° and are exposed in places where there is danger of loss of heat to surrounding objects or a cooling by evaporation and convection, must have provision made for maintaining the temperature. The loss of heat will depend upon the difference of temperature between the water in the tank and the earth or air outside and the conductivity of the walls, as in the case of heat losses from rooms, and the same values may be used. It is not likely, however, that such a refinement of calculation will be necessary, for the apparatus required for heating the water in the tank will prove of such large capacity that when operated at its slowest rate it will be sufficiently large to maintain the temperature of the water at the right degree.

To find the amount of heat required to heat the water in the tank the number of heat units necessary must be determined. This will depend upon the cubical quantity of water and the number of degrees it must be raised in temperature. It must be borne in mind that a heat unit is the quantity of heat required to raise the temperature of one pound of water one degree Fahrenheit—so to find the number of heat units the cubical contents of water must be reduced to pounds of water. This can be done by multiplying the number of cubic feet of water by 62.355; or if the contents are in gallons, by multiplying by 8.33, which is the weight of one gallon of water. Having the number of pounds of water, multiplying it by the raise in temperature will give the heat units required. For instance, if the water is to be raised in temperature from 55° to 75° Fahrenheit, a range of 20°, then multiplying the total number of pounds of water by 20 will give the heat units required.

The heat required to warm the water in a tank can be found by this rule:

173

RULE.—Multiply the weight of water in pounds by the rise in temperature of the water. The product will be the total heat units required.

EXPRESSED AS A FORMULA:

$$wt = h.$$

In which w = weight of water in pounds; t = degrees temperature water is to be raised; h = heat units required.

EXAMPLE.—A plunge bath is 30 feet long, 20 feet wide and has an average depth of water of 5 feet. How many heat units will be required to raise the temperature of the water from 50° to 75° Fahrenheit?

SOLUTION.—A volume of water 30 x 20 x 5 feet contains 3,000 cubic feet, or 62.355 × 3,000 = 187,-065 pounds. The difference between 75° and 50° = 25°. Substituting these values in the formula:

187,065 × 25 = 4,676,625 heat units. (Answer.)

Knowing the quantity of water to be heated, the next step is to find the size of apparatus to heat that quantity of water, or the capacities of water heaters and steam coils.

Capacity of Water Heaters.—The capacity of a water heater depends upon the amount of coal it can efficiently burn during a given period of time and the conductivity and thickness of the walls of the firebox. Boiler iron is a better conductor of heat than cast iron, therefore a boiler-iron heater of given surface will heat more water in an hour than will a cast-iron heater of equal surface, the amount of coal burned and the intensity of the fire in both cases being equal. The amount of coal economically burned in a heater depends upon the area of grate and size of the smoke flue. Heaters burn from three to six pounds and will probably average four pounds of coal per hour per square foot of grate surface. The total heat of combustion of a pound of coal of average composition is 14.133 B. T. U. Of this amount, however, a large percentage passes up the chimney as hot gases, so that under ordinary conditions only about 8,000 B. T. U. are actually transmitted to the water. Therefore, in calculating the capacity of a heater, the area of grate surface, amount of coal efficiently burned and the available B. T. U. in a pound of coal are the limiting factors. Architects and plumbers should determine for themselves, by calculation, the heating capacity of a heater, and not rely upon manufacturers' ratings. This is made necessary by the lack of uniformity among manufacturers in the rating of their heaters, which differ from one another in some cases over 100 per cent. for equal area of grates. Some part of that percentage might be accounted for by the difference of construction, which gives some heaters greater heating surface than others, but, making due allowance for the improved design of some heaters, they will invariably be found overrated, while the run of heaters are overrated from 20 to 50 per cent. The capacity of heaters can be calculated by means of the rule or formula following:

When the quantity of water to be heated per hour is known the size of grate required can be found by the following rule:

RULE.—Multiply the weight of water in pounds by the number of degrees' rise in temperature, and divide the product by the number of pounds of coal burned per hour per square foot of grate surface times the number of heat units transmitted to the water from one pound of coal. The result will be the area in square feet of grate required.

EXPRESSED AS A FORMULA:

$$g = wt \div cu.$$

In which w = weight in pounds of water to be heated; t = degrees Fahrenheit water is to be raised; c = pounds of coal burned per hour per square foot of grate; u = units of heat absorbed by water from each pound of coal; g = area of grate in feet.

EXAMPLE —What size of grate will be required to heat 300 gallons of water per hour from 62° to 212° Fahrenheit, 1 gallon weighing 8.3 pounds?

SOLUTION.—$\dfrac{300 \times 8.3 \times (212 - 62)}{6 \times 8,000} = 7.7$

square feet grate surface. (Answer.)

In the above solution six pounds of coal was assumed as the consumption per square foot of grate surface because the maximum rating of the heater is desired.

The capacity of a water heater of known dimensions can be ascertained by the following rule:

RULE.—Multiply the consumption of coal per square foot of grate surface by the number of B. T. U. transmitted to the water from each pound of coal, then by the number of square feet of surface in the grate, and divide the product by the weight of one gallon of water times the degrees of temperature the water is raised.

EXPRESSED AS A FORMULA:

$$q = gcw - pt.$$

In which g = size of grate in square feet; c = pounds of coal burned per hour per square foot of grate surface; u = units of heat absorbed by the water from each pound of coal; p = 8.3 weight of 1 pound of water; t = degrees Fahrenheit water is raised; q = quantity of water in gallons heated per hour.

EXAMPLE—How many gallons of water can be heated from 62° to 212° Fahrenheit in a heater with 7.7 square feet grate surface?

$$\text{SOLUTION.} - \frac{7.7 \times 6 \times 8,000}{8.3. \times (212 - 62)} = 296.8. \quad \text{(Ans.)}$$

Smoke Flues.—It is important that a good chimney flue, straight and smooth inside and proportioned to the area of the grate, be provided for each water heater. No other smoke pipe should be permitted to connect to this flue, nor should other openings to it be permitted, as they would spoil the chimney draught. Smoke flues should be cased with flue linings to give them a smooth interior surface. The best form for flue linings is round or oval, as smoke and hot gases pass up with less frictional resistance in a round flue than in a square one. Square flues are much more efficient than rectangular ones, on account of the less surface exposed for a given area of flue; for instance, a flue 12 x 12 inches has an area of 144 square inches and a perimeter of only 48 inches, while a flue 8 x 18 inches having an equal area has a perimeter of 52 inches, thus presenting four additional inches to offer resistance. No satisfactory formula was ever devised to calculate the area of smoke flues under varying conditions. A simple empirical rule that will be found satisfactory for determining the area of flues for water heaters follows:

RULE.—Allow for smoke flue one-eighth the sectional area of heater grate.

EXAMPLE.—What size of smoke flue will be required for a water heater containing 4 square feet of grate?

SOLUTION.—4 square feet = 576 square inches. $\frac{1}{8}$ of 576 = 72 square inches = area of smoke flue.

The nearest sizes of commercial flue linings are: square, $8\frac{1}{2} \times 8\frac{1}{2}$ inches = 72.25 square inches; round, $10 \times 10 \times .7584 = 78.54$ square inches.

Capacity of Steam Coils.—Steam coils for plunge baths and swimming pools are not advisable, as previously pointed out, although when necessary they may be used and may be made of copper, brass, or iron pipe, although brass or copper are the more suitable. Copper and brass pipes last longer than iron and transmit more heat to the water per square foot of heating surface, thereby requiring less room in the tank, and, besides, are easier to keep clean, as they do not rust. For these reasons either copper or brass coils are preferable to iron pipe coils. The size of steam coil in square feet required to heat a certain quantity of water in a given time can be found by the following rule:

RULE.—Multiply the weight of water in pounds by the number of degrees temperature Fahrenheit the water is to be raised and divide the product by the coefficient of transmission times the difference between the temperature of the steam and the average temperature of the water.

EXPRESSED AS A FORMULA:

$$s = wr - c (T - t).$$

In which s = surface of copper or iron pipe in square feet; w = weight in pounds of water to be heated; r = rise in temperature of water; t = average temperature of the water in contact with coils; T = temperature of steam; c = coefficient of transmission.

The value of c for copper is 300 B. T. U. and for iron 200 B. T. U. transmitted per hour per square foot of surface for each degree difference between the temperature of the steam and the average temperature of water.

In computing the heating surface of copper or iron pipe in steam coils the inner circumference of the pipe must be taken, as that is the real heating surface to which heat is applied. The average temperature of the water in contact with the coil is taken as the temperature of the water.

EXAMPLE.—How many square feet of heating surface will be required in a copper coil to heat 300 gallons of water per hour from 50° to 200° Fahrenheit with steam 15 pounds pressure?

SOLUTION.—$300 \times 8.3 = 2490$ pounds of water to be heated; $200° - 50° = 150° =$ rise in temper-

ature of water; $150° \div 2 + 50° = 125° =$ average
temperature of water; $250° =$ temperature of steam
at 15 pounds gauge pressure (Table II); $250° - 125°$
$=$ difference between temperature of steam and
average temperature of water. Substituting these
values in the formula:

$$s = \frac{2,490 \times 150}{300 \times (250 - 125)} = 9.9 \text{ square feet of coil.}$$

(Answer.)

Some convenient constants for steam coils that
produce approximations sufficiently accurate for
most purposes follow. The values will be found
safe:

$W =$ gallons water to heat per hour.

$W \div 10 =$ square feet iron pipe coil required
for exhaust steam.

$W \div 15 =$ square feet copper coil required for
exhaust steam.

$W \times .07 =$ square feet iron pipe coil for 5
pounds steam pressure.

$W \times .045 =$ square feet copper pipe coil for 5
pounds steam pressure.

$W \times .05 =$ square feet iron pipe coil for 25
pounds steam pressure.

$W \times .035 =$ square feet copper pipe coil for 25
pounds steam pressure.

$W \times .04 =$ square feet iron pipe coil for 50
pounds steam pressure.

$W \times .025 =$ square feet copper pipe coil for 50
pounds steam pressure.

$W \times .03 =$ square feet iron pipe coil for 75
pounds steam pressure.

$W \times .02 =$ square feet copper pipe coil for 75 pounds steam pressure.

Taking the foregoing example for comparison the nearest value to 15 pounds steam is 25 pounds, and the coefficient for copper pipe at this temperature is .035. Hence $300 \times .035 = 10.5$ square feet. (Answer.)

Heating Water by Steam in Contact.— The quickest and most economical way to heat water with steam is to bring the steam into direct contact with the water. This method is used extensively to heat water in swimming pools where noise is nòt objectionable, vats for industrial purposes, dish washing, etc., and is usually accomplished by forcing steam through a perforated pipe or steam nozzle located near the bottom of the tank and submerged by the water. When perforated pipes are used for this purpose they should be of brass or copper to prevent corrosion, and the combined area of the perforations should be at least eight times the area of pipe to equal it in capacity. Exhaust steam from pumps, engines or other apparatus that is liable to contain oil or grease is not suitable for this purpose.

When steam is brought in contact with water in an open vessel steam bubbles are formed, rise toward the surface and collapse with a report. For this reason water is heated by steam in direct contact through perforated pipes only when noise is not objectionable.

Steam Required to Heat Water.— The weight of steam required to heat a given quantity of water when brought into contact with it can be found by the following rule:

RULE. Multiply the number of pounds of water to be heated by the number of degrees temperature the water is to be raised, and divide the product by the total heat of steam at the pressure it is to be used, less the sensible heat at atmospheric pressure.

EXPRESSED AS A FORMULA:

$$s = wh \div (L-1)$$

In which s = weight of steam in pounds; w = pounds of water to be heated; h = degrees Fahrenheit water is to be heated; L = total heat of steam at pressure used; l = sensible heat at atmospheric pressure.

EXAMPLE.—How many pounds of steam at 70 pounds pressure will be required to heat 7,500 pounds of water from 48° Fahrenheit to boiling point?

SOLUTION.—$\dfrac{7,500 \times (212-48)}{(1,174-180)} = 1,237$ pounds of steam. (Answer.)

An empirical rule that is sufficiently approximate for most purposes is to allow one pound of steam for six pounds of water to be heated. Taking the above example then $7,500 \div 6 = 1,250$ pounds of steam. (Answer.)

The temperature of steam at different pressures can be found in Table II. To use this table add 14.7 to the reading of the gauge to get absolute pressure; the quantities desired will be opposite this figure.

CHAPTER XII

EXAMPLE OF A TURKISH BATH

S O far, the design of the Turkish bath has been treated mainly from an engineering standpoint, no consideration having been given to the architectural features of the building interior. Indeed, little can be said on that subject, for the architectural treatment must of necessity differ with the amount at the disposal of the designer.

The best way, perhaps, to treat the matter of architectural design, is to show views of the interior of one of the finest Turkish bath buildings in the country, the Fleishman Baths, of New York, designed by Buchman and Fox of that city. The entrance to the dressing rooms is shown in Fig. 42. There are 200 well ventilated dressing rooms in the group, each provided with a large comfortable divan, for sleeping or resting after the bath. From the dressing room the bathers go to the first hot room, or tepidarium, shown in Fig. 43, which is kept at a temperature of 150 degrees Fahrenheit. From this room, those who require a greater heat can pass to the second hot room, or calidarium, where the temperature is maintained at 180 degrees Fahrenheit.   A stay of from ten to fifteen minutes in these rooms is generally sufficient for persons who are not taking the baths for remedial purposes. From the dry hot rooms the bather can pass to the vapor room, Fig. 44, if he wishes, or those who

wish to take a Russian bath without entering the hot rooms start here in their course of treatment. This room is kept filled with steam vapor at a temperature of 112 degrees Fahrenheit, which is about as high a moist temperature as the average person can stand with comfort.

184

In the shampooing rooms, Fig. 45, the bathers are washed and scrubbed and kneaded, until all the loose and dead skin or epidermas hanging to the body is removed and washed away. Cold and hot water, and sea salt if it is wanted, combine to clean the hair and body, and cleanse the little pores of the skin of all foreign matter. There are four-

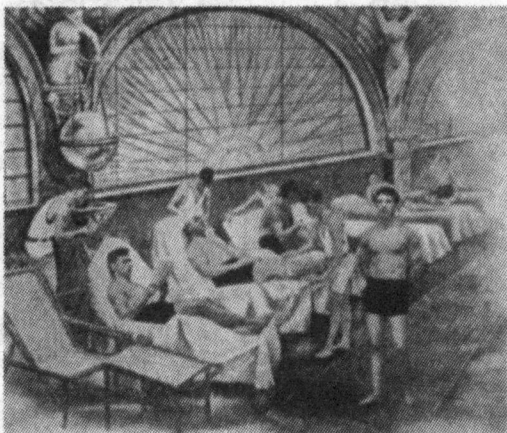

Fig. 43
The Calidarium, Hot Rooms

teen of these little rooms, and in this department are also the various needle, rain, douche, sitz and other scientific bath equipments.

A douche bath, and the manner of using it also, are shown in Fig. 46. The operator standing at his control table, about twenty feet from the bather, can throw a stream of water on various parts of the bather's body. By manipulation of

185

Fig. 44
Steam or Vapor Room

the valves, the water can be made as cold or as hot as the bather can stand, or can be moderated to a comfortable temperature, and the water can be made to strike him in a heavy stream, or in a soft, delightful, spray. The force of the water, together with the varying temperatures, makes this a very beneficial form of bath in many nervous complaints. The force of the water can be varied

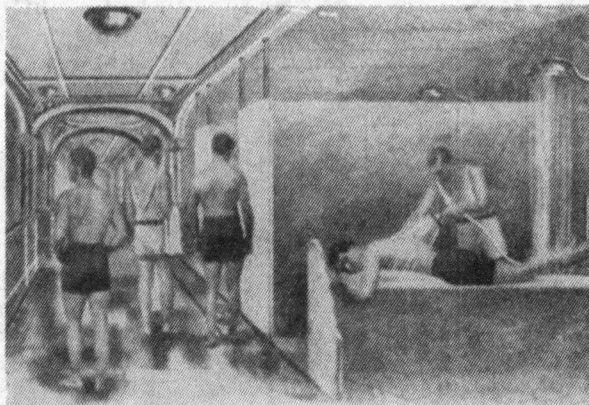

by the operator so that it will strike the bather with a pressure of from a few pounds to sixty pounds per square inch.

One of the delights of a Turkish bath is the cold plunge following the shampoo or scrubbing. In Fig. 47 is shown the plunge bath during the women's hour, and the rich architectural treatment of the compartment speaks for itself. The plunge

187

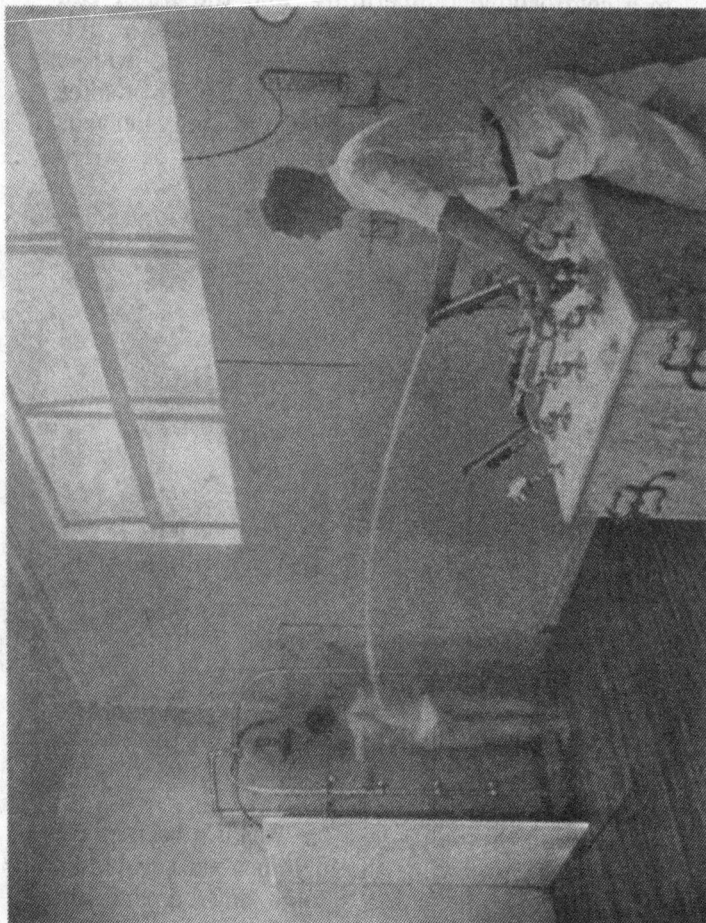

Fig. 46
Douche Bath

Fig. 47
The Plunge Bath

bath is on one of the upper floors of the building, and of necessity is a hanging bath, supported on heavy I beams. The water in this tank is kept fresh and pure by constantly renewing it, the entire contents being changed every half hour.

The last stage of a Russian or Turkish bath, and a very important one when rightly performed, is the massage. The compartment set aside for

this purpose is shown in Fig. 48. Here all the muscles of the body are worked and kneaded until they are loose and pliable, and in splendid working condition. It is here and at this time, too, that the body is rubbed with essential oils, to soften and keep pliable the skin, and prevent too rapid evaporation from the body until such time as the

190

natural oil glands of the body have a chance to resume their interrupted function.

The pedicure department is shown in Fig. 49. In addition to this, there is a barber shop ; manicure department; an electric vibratory department where vibratory massage, faradic, galvanic and sinusoidal electric treatments are administered by electro therapeutic specialists; gymnasium and electric light cabinets.

There are one hundred 16 candle power incandescent electric lights in each bath cabinet, and the numerous rays from these lights are multiplied and focussed on the occupant of the cabinet by mirrors and reflectors. In addition to the gentle rays of heat given off by the electric light bulbs, the direct and reflected rays of light concentrated on the body penetrate the tissues thereby stimulating the vital forces.

Fig. 80
The Solarium

A feature of this Turkish bath is the Solarium on the top floor of the building, covered or roofed over with glass so the patrons can enjoy the benefit and luxury of a sun bath. A well equipped gymnasium affords an opportunity for exercise, either before or after the bath, while private sleeping rooms, club, café and grill rooms, complete the list of compartments in this very interesting building, which may well serve as a guide in designing similar buildings.

On account of their beneficial action in maintaining health and curing diseases, Turkish baths ought not to be restricted to these public or private enterprises found in all cities, but ought by right to become regular parts of all hospitals, hotels, public bath houses, paid baths, asylums, sanitoriums, clubs, Young Men's Christian Association Buildings, Young Women's Christian Association Buildings, orphan homes, barracks, poor houses, homes for the aged, and private homes. A little sweat room can be fitted up so easily and with such little cost, that no home ought to be without one. The money expended in this way will save itself over and over again in the doctor bills avoided, and in the increased efficiency attendent on good health.

INDEX

♨ ♨